SAFETY AND HEALTH MANAGEMENT IN THE NINETIES

SAFETY AND HEALTH MANAGEMENT IN THE NINETIES

Creating a Winning Program

Milton J. Terrell

VNR VAN NOSTRAND REINHOLD
I(T)P A Division of International Thomson Publishing Inc.

New York • Albany • Bonn • Boston • Detroit • London • Madrid • Melbourne
Mexico City • Paris • San Francisco • Singapore • Tokyo • Toronto

I(T)P™ A division of International Thomson Publishing Inc.
The ITP logo is a trademark under license.

Printed in the United States of America
For more information contact:

Van Nostrand Reinhold
115 Fifth Avenue
New York, NY 10003

International Thomson Publishing GmbH
Königswinterer Str. 418
53227 Bonn
Germany

International Thomson Publishing Europe
Berkshire House,168-173
High Holborn, London WC1V 7AA
England

International Thomson Publishing Asia
221 Henderson Road #05-10
Henderson Building.
Singapore 0315

Thomas Nelson Australia
102 Dodds Street
South Melbourne 3205
Victoria, Australia

International Thomson Publishing Japan
Hirakawacho Kyowa Building, 3F
2-2-1 Hirakawacho
Chiyoda-ku, 102 Tokyo
Japan

Nelson Canada
1120 Birchmount Road
Scarborough, Ontario
Canada M1K 5G4

International Thomson Editores
Campos Eliseos 385, Piso 7
Col. Polanco
11560 Mexico D.F. Mexico

1 2 3 4 5 6 7 8 9 10 BBR 01 00 99 98 97 96 95

Library of Congress Cataloging-in-Publication Data

Terrell, Milton J.

Safety and health management in the nineties: creating a winning program / Milton J. Terrell.

p. cm.

Includes index.

ISBN 0-442-02056-2

1. Industrial safety--Management. 2. Industrial hygiene--Management. I Title.

T55.T43 1995

658.4' 08--dc20 95-6258

CIP

This book is dedicated to Peggy Richardson,
whose continued efforts have made the VPP
what it is today and will be in the future.

Contents

Introduction

As we close this century, safety and health management is in a stage of transition. Companies have recognized this and modernized their programs. Many of these companies have entered the Occupational Safety and Health Administration's (OSHA) "Voluntary Protection Program" (VPP). This book has been compiled from materials in the safety and health management guidelines developed by OSHA and from a set of pamphlets drafted by the staff of the VPP. Using the material presented in this book will give any company the ability to modernize its program to meet the safety and health program requirements in the new century.

In the early 1980s, the Occupational Safety and Health Administration developed the Voluntary Protection Program. This program was a step away from the traditional compliance-oriented role of OSHA. The VPP emphasized safety and health program management rather than strict compliance to a set of rigid standards. Responsibility and accountability for all employees are an important part of the VPP, as are intensive employee participation in all aspects of the safety and health program. A set of criteria was developed to measure an applicant's suitability for participation in the VPP. From these criteria, OSHA produced the "Safety and Health Management Guidelines." These guidelines were not mandatory and have not been a part of the OSHA compliance effort, which remains staunchly compliance-oriented. The Safety and Health Management Guidelines, in a condensed version, follow.

Management Leadership and Employee Involvement

- A clearly stated policy so that all employees understand the priority of safety and health concerns
- Goals and objectives for the program
- Visible top management involvement in safety and health
- Employee involvement in the structure and operation of the program
- Assignment and communication of responsibility so that all managers, supervisors, and employees in all parts of the organization know what performance is expected of them
- Adequate authority and resources to carry out assigned responsibilities
- Accountability for all managers, supervisors, and employees in meeting their assigned responsibilities
- Review of program operations at least annually to evaluate their success in meeting the goal and objectives of the overall program

Worksite Analysis

- Identification of all hazards through comprehensive surveys; analysis of change to facilities, equipment, materials, and processes; and routine hazard analysis
- Regular site safety and health inspections
- Reliable systems for employees to report hazards
- Investigation of accidents and near misses
- Analysis of the patterns of injuries and illnesses

Hazard Prevention and Control

- Use of engineering controls, work practices, personal protective equipment, and administrative controls
- Preventive maintenance
- Planning and preparation for emergencies
- A medical program

Safety and Health Training

- Ensuring that all employees understand hazards and how to prevent harm to themselves and others
- Ensuring that supervisors understand their safety and health responsibilities to identify previously unrecognized hazards, maintain physical protections, and reinforce employee training through feedback on performance and enforcement when necessary
- Ensuring that managers understand their safety and health responsibilities

Sometime in the late 1980s, members of the VPP staff undertook the task of putting together a set of pamphlets based on the Safety and Health Management Guidelines.

Developed to aid VPP participants and potential participants in bettering their safety and health programs, the pamphlets as a whole took modern ideas regarding safety and health programs and added aspects of quality management theory. The pamphlets could have been an excellent source of information for any type of industry, but for various reasons were not widely distributed.

This book takes the information from these pamphlets and puts it into a single source. In this form, it can be used as a tool for developing a safety and health program or as an improvement guide for a program already in place. Proposed federal legislation would make written safety and health programs mandatory for most of industries. The guidelines have been recommended as the blueprint for these mandatory programs. Several states, including Texas and Nevada, have made portions of the guidelines part of mandatory written safety and health programs for all industries. These requirements are addressed directly in this book.

It is the intent of this book to present a guide for companies—particularly those that do not have full-time safety and health professionals on their staff—that would allow them to develop a safety and health program of VPP quality, with only limited assistance from a safety and health professional.

It has been my experience that most companies view safety and health as an entity that does not integrate into their normal way of doing business. Many companies spend large amounts of money and resources for a quality management program, but do not consider safety and health a part of that program. Expenses for safety and health, workers' compensation insurance, and OSHA penalties are generally considered part of the cost of doing business. The participants in the VPP have shown that integrating safety and health into the total business picture can save large amounts of money in reduced workers' compensation costs and removal of major OSHA penalties. Most of the companies showed an increase in productivity, and none showed a decrease. These savings are all hard dollars that go directly onto the bottom line as profit.

If I were asked which aspects of this book are the most important for a successful safety and health program, I would have to identify three:

- Responsibility
- Accountability
- Employee participation

These are the foundation for any modern safety and health program. The chapters covering these aspects clearly explain the need for each and how to implement them into a safety and health program.

Acknowledgment

This book was written using, in most part, a series of unpublished pamphlets from the U.S. Department of Labor, Occupational Safety and Health Administration, Voluntary Protection Program. These pamphlets contained some of the best information that I had ever seen from any governmental agency, and I felt that it needed much broader dissemination. My thanks to those individuals with the VPP who wrote the drafts for the pamphlets.

1

Safety and Health Management Guidelines

You *can* do a better job of protecting your workers from the hazards in worksites large and small. Better management will produce lower costs, higher productivity, and better employee morale. No matter how small your worksite, you can approach worker protection with systematic methods that will work for you.

Effective management of worker safety and health protection is crucial in reducing the numbers and severity of workplace illnesses and injuries. To be effective, such management must include the methods to find and understand all hazards, including potential hazards that could result from a change or series of changes in conditions or practices. Then management must either prevent or control those hazards so that workers are not exposed.

The Safety and Health Management Guidelines were developed from the OSHA-VPP requirements and the safety and health program elements of other safety and health programs. Managers are encouraged to implement programs that meet these guidelines in order to provide good worker protection from hazards.

The Guidelines

The guidelines outline a management program to provide for the recognition and understanding of all the hazards and potential hazards of the workplace, the prevention or control of those hazards, and the training to ensure that employees at all levels understand the potential hazards they may be exposed to and how to help protect themselves and others. To accomplish this, the guidelines are divided into four major elements:

1. Management Leadership and Employee Involvement
2. Worksite Analysis
3. Hazard Prevention and Control
4. Safety and Health Training

Each of these elements is further divided into several recommended actions.

Management Leadership and Employee Involvement

This element describes the leadership that management provides to encourage everybody at the worksite to be involved in safety and health protection. Many of the actions listed under this element are measures that apply to all phases of business management. The guidelines simply put them to use in improving worker safety and health protection. The actions cover the following:

1. Safety and Health Policy
2. Goal and Objectives
3. Visible Top Management Involvement
4. Employee Involvement
5. Assignment of Responsibility
6. Provision of Adequate Authority and Responsibility
7. Accountability
8. Program Evaluation

These actions lead from one to another in a logical flow, except that a special emphasis on top management and employee involvement flows from assignment of responsibilities. The flow of logic can be diagrammed as follows:

Policy → Goal → Objectives → Assignment of Responsibilities →
↓ Visible Management Involvement
↓ Employee Involvement

→ Provision of Authority and Resources → Accountability → Program Evaluation

Safety and Health Policy

Developing a clear statement of management policy helps everyone involved with the worksite understand the importance of safety and health protection in relation to other organizational values. Clearly communicating the policy to all employees ensures that no confusion will exist when a conflict arises between two of these values, such as productivity and safety or health. The guidelines describe the desired action as follows:

> State clearly a worksite policy on safe and healthful work and working conditions, so that all personnel with responsibility at the site and personnel at other locations with responsibility for the site understand the priority of safety and health protection in relation to other organizational values.

Goal and Objectives

By establishing a clear goal and objectives for the safety and health program, management makes the general policy specific. The goal and objectives set the framework for assigning responsibility. Each employee should be able to see his or her work activities in terms of moving toward the goal and achieving objectives. The guideline is as follows:

> Establish and communicate a goal for the safety and health program and objectives for meeting that goal, so that all members of the organization understand the results desired and the measures planned for achieving them.

Visible Top Management Involvement

If employees can see the emphasis that top management puts on safety and health and the leadership provided in making the program work, they are more likely to emphasize it in their own activities. It is important for worksite managers to follow set safety and health rules and work practices scrupulously to provide an example that hourly workers will follow. Management leadership should include ensuring equal safety and health protection of any contract workers at the site. Worksite managers should also show their visible involvement through such means as making plantwide safety and health inspections; chairing the safety and health committee; personally stopping hazardous activities or conditions until the hazards can be corrected or controlled; and/or personally tracking safety and health performance and holding managers and employees accountable. Remember, actions speak louder than words. The guideline is as follows:

> Provide visible top management leadership in implementing the program and ensuring that all workers, including contract workers, are provided equally high quality safety and health protection, so that all will understand that management's commitment is serious.

Employee Involvement

The best worker safety and health protection occurs where everyone at the worksite feels responsible for that protection. To make that happen, all employees must know that they are helping to shape the program. Employees at all levels should be actively involved in finding and correcting safety and health problems. This does not mean that the employer gives up responsibility and authority. The Occupational Safety and Health Act places clear responsibility for worker protection from occupational hazards squarely on management. The wise manager, however, uses employees' unique knowledge and experience to help find problems and get them resolved successfully. The guideline is as follows:

> Provide for and encourage employee involvement in the structure and operation of the program and in decisions that affect their safety and health, so that they will commit their insight and energy to achieving the safety and health program's goal and objectives.

Assignment of Responsibility

Everyone in the workplace should have some responsibility for safety and health. Clear assignment helps avoid overlap or gaps in getting the needed activities done. In particular, you should make sure that the safety and health "expert" at the worksite is not assigned line responsibility that belongs to line managers and supervisors, such as supervising and evaluating workers' performance in areas of safety and health, providing on-the-job training in safe work practices and use of personal protective equipment, or encouraging workers' participation in safety and health activities.

The responsibilities should flow logically from the objectives set to meet the overall goal of the safety and health program. The guideline is as follows:

> Assign and communicate responsibility for all aspects of the program, so that managers, supervisors, and employees in all parts of the organization know what performance is expected of them.

Provision of Adequate Authority and Responsibility

Realistic assignment of responsibility must include provision of appropriate authority and adequate resources, including appropriately trained and equipped personnel as well as sufficient operational and capital funding. The guideline is as follows:

> Provide adequate authority and resources to responsible parties, so that assigned responsibilities can be met.

Accountability

Once an individual has been assigned responsibility and given the appropriate authority and resources, management must follow up and hold that person accountable for achieving what he or she has been asked to do. Accountability is crucial to help each person understand how critical his or her performance is, and to teach them to take personal responsibility for their performance. The guideline is as follows:

> Hold managers, supervisors, and employees accountable for meeting their responsibilities, so that essential tasks will be performed.

Program Evaluation

Once the program is implemented and operating, it is important to ensure the quality of results by evaluating them against the goal and objectives that were supposed to be achieved. This should be performed with these questions in mind: "Did we get where we wanted to go?" and "Did each specific activity help us get there?" The guideline is as follows:

> Review program operations at least annually to evaluate their success in meeting the goal and objectives, so that deficiencies can be identified and the program and/or the objectives can be revised when they do not meet the goal of effective safety and health protection.

Worksite Analysis

Worksite analysis is a combination of systematic actions that provide you with the information you need to recognize and understand the hazards and potential hazards of your workplace. While they may look complicated at first glance, they consist of activities that already are being performed in most workplaces. In the guidelines they are differentiated from each other for the sake of clarity. The actions are as follows:

1. Comprehensive Hazard Identification
    a. Comprehensive hazard surveys
    b. Hazard analysis of changes in the workplace
    c. Routine hazard analysis
2. Regular Site Safety and Health Inspections
3. Employee Reports of Hazards
4. Accident/Incident Investigations
5. Injury and Illness Trend Analysis

Comprehensive Hazard Identification

There are three aspects of a complete hazard inventory from which a program of prevention and control can be designed. The first of these is the comprehensive hazard survey. This is the most basic of all the tools used to establish the inventory of hazards and potential hazards at your worksite. This survey is best performed by experts from outside the worksite who have knowledge that is both broad and deep. That expertise should include safety engineering, industrial hygiene, and, in most cases, occupational "medicine." After a baseline has been established, only periodic comprehensive surveys need be done to take advantage of new information in the experts' "community" about newly understood hazards or means of controlling them.

The second aspect of comprehensive hazard analysis is **change analysis**, which means precisely what the name implies. Each time there is a change of facilities, equipment, processes, or materials in your workplace, they should be analyzed for hazards *before they are introduced.* This way, not only is harm to workers avoided, but the extra expense of retrofitting controls after installation and use is also avoided.

The last aspect of a complete hazard inventory is **rou-**

tine hazard analysis. The basic form of this analysis, which is useful at every type of worksite, is the job safety analysis (JSA). This analysis breaks a job into tasks and steps, analyzes the hazards that are potential in each step, and then provides a method of prevention or control to reduce exposure. A variation of this, used at worksites with highly complex hazards such as chemicals or nuclear energy, is the **process hazards analysis.** This analysis breaks a process down to its smallest components so that hazards can be identified and preventions or controls devised. In rapidly changing workplaces such as construction, **activity hazard analysis** is a useful form of the routine hazard analysis. In activity hazard analysis, each phase of the rapidly changing work is analyzed for the new hazards it may introduce so that preventions or controls can be devised.

The guideline is as follows:

So that all hazards are identified:

Conduct comprehensive worksite surveys to establish safety and health hazard inventories and update them periodically as expert understanding of hazards and the methods of control in your industry change;

Analyze planned and new facilities, processes, materials, and equipment; and,

Perform routine hazard analysis of jobs, processes and/or varied phases of work, as needed.

Regular Site Safety and Health Inspections

General site inspections should be performed by personnel at the worksite who have been trained in recognizing hazards that have a tendency to "slip out" of the controls designed to reduce employee exposure to them. Inspectors should also keep an eye out for hazards that may not have been identified in the comprehensive survey or through other means of establishing the hazard inventory. The guideline is as follows:

> Provide for regular site safety and health inspections, so that new or previously missed hazards and failures in hazard controls are identified.

Employee Reports of Hazards

By using every employee as an extra pair of eyes to spot hazards, you can greatly increase your speed and reliability in finding and correcting problems before any harm is done. Providing one or more systems for employees to alert you to hazards should be accompanied by protection of those employees from any harassment that might result from such a report. Employees also need to see timely and appropriate responses to their reports to encourage them to continue their help. Such responses also provide employees with evidence of management's commitment to the safety and health program. The guideline is as follows:

> So that employee insight and experience in safety and health protection may be utilized and employee concerns addressed, provide a reliable system for employees, without fear of reprisal, to notify management personnel about conditions that appear hazardous and to receive timely and appropriate responses; and encourage the employees to use the system.

Accident/Incident Investigation

Investigating accidents and incidents (without injuries) is another opportunity to find hazards and design prevention and/or controls. For each accident, there are usually several steps to be taken to prevent future occurrences. The guideline is as follows:

> Provide for investigation of accidents and "near miss" incidents, so that their causes and means for preventing their repetitions are identified.

Injury and Illness Trend Analysis

Over a period of time, a review of injuries and illnesses—including illness that are apparently not occupationally related—may reveal patterns or clusters that indicate common worksite causes or origins that did not become apparent as each individual case was addressed. The guideline is as follows:

> Analyze injury and illness trends over time, so that patterns with common causes can be identified and prevented.

Hazard Prevention and Control

Once the hazards and potential hazards of the workplace have been inventoried, a program of prevention and control can be designed. This consists of the following actions:

1. Establishing Appropriate Controls
   a. Engineering controls where feasible
   b. Safe work practices with enforcement, where necessary, through a disciplinary system
   c. Personal protective equipment (PPE)
   d. Administrative controls
2. Providing Preventive Maintenance
3. Preparing for Emergencies
4. Implementing a Medical Program

Establishing Appropriate Controls

In designing a program of prevention and control, prevention of exposure to a hazard is always the best choice wherever feasible. This means removing the hazard or preventing exposure through engineering controls. Where complete removal is not feasible, the next best choice is complete enclosure. Where complete enclosure is not feasible, a combination of partial enclosure and work practices, perhaps including PPE, is the next best choice. Where no enclosure is possible, a combination of work practices and PPE should be used where feasible.

Please remember that work practices and PPE place special responsibilities on the employees who use them. Employees should be trained to understand why these protections are necessary and how they can use these methods to protect themselves and others. Management should stress in every possible way the importance of these protections, including, when necessary, the use of fair and consistent discipline.

When all other controls fail to provide enough reduction in exposures, administrative controls, such as worker rotation, should be used. The guideline is as follows:

> So that all current and potential hazards, however detected, are corrected or controlled in a timely manner, establish procedures for that purpose, using the following measures:
>
> Engineering techniques where feasible and appropriate;
>
> Procedures for safe work which are understood and followed by all affected parties, as a result of training, positive reinforcement, correction of unsafe performance, and, if necessary, enforcement through a clearly communicated disciplinary system;
>
> Provision of personal protective equipment; and,
>
> Administrative controls, such as reducing the duration of exposure.

Providing Preventive Maintenance

A good equipment maintenance program can keep engineering control systems functioning as intended and prevent ordinary nonhazardous equipment from becoming hazardous. The guideline is as follows:

> Provide for facility and equipment maintenance, so that hazardous breakdown is prevented.

Preparing for Emergencies

You should plan and prepare for all possible emergencies. The more likely the possibility of an emergency, the more preparation should be done. All employees should know exactly what they must do in each type of emergency. They should practice it enough so that it becomes easy to do in stressful emergencies. The guideline is as follows:

> Plan and prepare for emergencies, and conduct training drills as needed, so that the response of all parties to emergencies will be "second nature."

Implementing a Medical Program

Having a medical program onsite does not necessarily mean having a doctor, or even a nurse, onsite. It does mean involving occupational health professionals in worksite analysis for hazards, in hazard prevention and control programs, in early recognition and treatment of injuries and illnesses, and in limiting the severity of illness and injury. In addition to health professionals, other employees at the site should be trained in first aid and CPR. The guideline is as follows:

> Establish a medical program which uses occupational health professionals in the analysis of hazards, early recognition and treatment of illness and injury, and

the limitation of the severity of harm; and which provides first aid and cardiopulmonary resuscitation (CPR) onsite and physician and emergency medical care nearby, so that harm will be minimized if an injury or illness does occur.

Safety and Health Training

In a good program of safety and health management, it is crucial that all people at the worksite understand their role in that program, the hazards and potential hazards that need to be prevented or controlled, and how to protect themselves and others. The actions to achieve this are as follows:

1. Ensure that **employees** understand hazards
2. Ensure that **supervisors** understand their responsibility for:
   a. Analyzing the work under their supervision for hazards
   b. Maintaining physical protections
   c. Reinforcing and enforcing needed protective measures
3. Ensure that **managers** understand their responsibilities

Employees

At a minimum, employees must know the general safety and health rules of the worksite, specific hazards and the safe work practices needed to help control exposure, and the employee's role in all types of emergency situations. This is usually achieved through thorough orientation, periodic safety and health training, and emergency drills. Special training may also be needed for the particular skills required for the job or for activities under the safety and health program. The guideline is as follows:

> Ensure that all employees understand the hazards to which they may be exposed and how to prevent harm to themselves and others from exposure to these hazards, so that employees accept and follow established safety and health protections.

Supervisors

Special training should be provided to supervisors to help them in their leadership role. They should be taught to look for hidden hazards in the work under their supervision, to insist upon the maintenance of the physical protections for work under their supervision, and to reinforce employee hazard training through performance feedback and, when necessary, fair and consistent enforcement. The guideline is as follows:

> So that supervisors will carry out their safety and health responsibilities effectively, ensure that they understand those responsibilities and the reasons for them, including:
>
> Analyzing the work under their supervision to identify unrecognized potential hazards;
>
> Maintaining physical protections in their work areas; and,
>
> Reinforcing employee training on the nature of potential hazards in their work and on needed protective measures, through continual performance feedback and, if necessary, through enforcement of safe work practices.

Managers

All line managers must understand their own responsibilities for safety and health. This will probably not require special classroom training but needs to be communicated clearly to all managers. The guideline is as follows:

> Ensure that managers understand their safety and health responsibilities as described under "Management Leadership and Employee Involvement," so that managers will effectively carry out those responsibilities.

Summary

Based on a variety of experiences, we are convinced that good management of worker safety and health protection will result in lower injuries and illnesses. Effective management of such protection will result in better employee morale, higher productivity, and improved product quality. This book can help you implement a quality management program to provide that protection. We believe your efforts to protect your workers with this type of safety and health program will be amply rewarded.

2 *Determining the Direction of Your Program: Establishing Policy, Goals, and Objectives*

When you plan a journey, you usually have a reason for going (policy), a destination (goal), and a direction (objective). Similarly, when planning your safety and health program, you first decide and state in writing your policy regarding safety and health. Then you decide upon a goal. Next you set the objectives that will help you reach that goal. Thus you have determined the direction of your program.

Your whole journey to good worker safety and health protection has been mapped out for you in the Safety and Health Management Guidelines. This chapter will help you start your journey to a good safety and health program by explaining how to write and communicate your safety and health policy, and how to set and evaluate your goal and objectives. You will find many examples and worksheets to help you on your way.

Policy

Every successful safety and health program has an active and aggressive commitment from top management to employee safety and health. This commitment in turn influences the action of the company's managers, supervisors, foremen, and employees, and will ultimately determine the effectiveness of the safety and health program in reducing or eliminating workplace injuries and illnesses. The company states its commitment through its written and clearly communicated policy for workplace safety and health, which makes clear the priority of safety and health in case of conflict with other priorities. This policy statement should be signed by the highest ranking company official on the site.

The Priority of Safety and Health

Successful companies consider workplace safety and health to be even more important than such priorities as production, sales, and quality control. If the policy statement makes that order of priorities clear, all employees will know what choice to make if a conflict between employee well-being and one of the other priorities arises. Some examples of policy statements that convey this belief are "People are our most important resource and their safety and health is our principal responsibility," "Every employee is entitled to a safe and healthful place in which to work," "No job is so important it can't be done in a safe and healthful manner," and "If it is not safe and healthful, we will not do it."

Communicating Your Policy

For your safety and health policy to be effective, it is critical to communicate it to all your employees. The policy expresses your attitude toward employee safety and health, and communicating it effectively signifies that your attitude is strongly positive. You must communicate your policy by *word, action,* and *example.* It will help to explain it both one on one and in group meetings. For larger workplaces, it will be necessary to communicate it in *writing.*

Through Explanation

New employees start learning about the company's attitude toward employee safety and health when they first start their jobs. Discussing the hazards of the job and providing training on safe work procedures tell the employee that safety and health has a high priority in your company. The supervisor's continued emphasis on safety and health reinforces this positive company attitude.

In Writing

A carefully written policy statement that clearly and simply conveys your attitude about safety and health is always recommended. In addition, communicating that attitude through your actions, example, and explanations will emphasize this commitment.

Except in very small companies where the policy is easily explained and understood through spoken statements, a written policy statement is vital. A written statement clarifies policy, creates consistency and continuity, serves as a checkpoint whenever there is a question of priority between safety and health and production, and gives supervisors support in the enforcement of safety and health rules and safe work practices.

You will want to include the written statement in the information you give new employees, and post it on employee information bulletin boards. Type it up on company letterhead, and be sure to sign it.

Keep in mind that the written statement of policy is *not* the policy, but is only one way of communicating the policy. The *real* policy is the attitude and concern that you show for the safety and health of your employees.

Appendix 1 is a policy statement worksheet, which provides a number of examples to show how the various portions of a policy statement can be written.

By Action

What we do speaks louder than what we say. Your commitment of resources to the prevention and control of unsafe or unhealthful work or working conditions, to the provision of safe work practices and personal protective equipment (PPE) where needed, and to safety and health training tells your employees, in positive ways, that you have a safety and health policy, and that your policy reflects real concern for their safety and health. Louder than that, your willingness to put safety and health before short-term productivity will speak clearly to your employees that you mean what your policy says.

By Example

Top management, supervisors, and foremen all express the company's attitude toward workplace safety and health by the daily example they set. Rules and regulations posted on bulletin boards and discussed in meetings can become a failure or a joke when management does not follow and enforce them. You can set an example by such actions as the proper use of PPE, the safe operation of equipment, holding supervisors accountable for their safety and health responsibilities, and running your business in a safe and healthful manner.

Goal

Now that you have determined the reason for your journey, you have come to the point of choosing your destination, determining where you want to be in your safety and health program. For your program to meet the intent of your safety and health policy statement, you need to identify and set the goal that you wish to achieve. To do this you need to think about where your policy says you want to be.

The policy statements discussed above all boil down to the same concept of desiring to provide work and working conditions that are not harmful to your employees. These statements also reflect the stated purpose of the Occupational Safety and Health Act of 1970 (29 U.S.C. 651 *et seq.*): "to assure so far as possible . . . safe and healthful working conditions." Section 5.(a)(1) of that act requires that each employer "furnish to each of his employees employment and a place of employment which are free from recognized hazards."

In determining what your goal will be, there are at least two types you will want to consider: numerical and descriptive.

Numerical Goal

Many employers set a numerical goal because it is easy to measure. The problem is that it is difficult to set a numerical goal that covers all aspects of the concept of providing a place of work and working conditions that are free from existing recognized hazards.

Zero Hazards

If you set, for example, a goal of no existing hazards at any time, it may be so difficult to reach that you and your employees may well become disillusioned long before you have a chance to reach that goal.

Certain Number of Injuries

You could set a goal for a certain number of injuries. In doing so, however, you will have left out illnesses and the existing hazards that did not result in an injury. The problems with this become important when you set your objectives. The objectives you set to help you ensure a low number of injuries will not necessarily be the same as those you would set to ensure a low number of illnesses.

Certain Number of Injuries and Illnesses

Adding illnesses to your goal may not help because illnesses are not always easy to recognize until long after the exposure to hazards that could have been prevented or better controlled. Even with a goal that includes a numerical reduction in both injuries and illnesses, you may not be setting objectives that will help you reduce exposure to hazards that have not yet resulted in injury or illness.

Descriptive Goal

If you set a goal to reduce exposure to hazards, it will be helpful in setting your objectives and easy to evaluate your results, but not as easy as a simple numerical goal. The most inclusive goal (which is attainable) is not numerical. This goal is to achieve a comprehensive program that assesses all hazards and known potential hazards of your worksite and prevents or controls those hazards. While not as succinct or as easily measurable as a numerical goal, this goal is attainable, and it is not difficult to evaluate your objectives and program results against such a goal. There may be other ways of stating this or similar concepts for your workplace, but OSHA recommends this as the best goal.

Objectives

In our analogy of a journey, after you establish the reason for the journey (policy) and the destination for the journey (goal), you are ready to determine the direction (objectives). These are the road signs to effective safety and health management. By setting objectives you will improve the odds of attaining a desired performance. To develop meaningful objectives, you need to:

1. Determine where you want to be
2. Determine where you are
3. Determine what you need to get where you want to be

In setting your goal you have determined where you want to be. Next you should determine where you are now.

Determine Where You Are

Before you establish your objectives, you need to gather as much information as possible about the current conditions at your workplace and about practices that are already a part of your safety and health program.

Is Your Safety and Health Program Complete?

At a minimum, your program should reflect these four basic elements:

1. Management Leadership and Employee Participation
2. Worksite Analysis
3. Hazard Prevention and Control
4. Safety and Health Training

If you do not yet have a safety and health program that includes all these elements, you will need to look at your workplace to decide what is missing or what needs improvement. You should consider developing objectives relating to the missing or incomplete action items in each element. This will help you create the best safety and health program you can develop for your worksite.

Get Everyone Involved

Here is an opportunity to get employees involved. Ask employees and supervisors to help you identify both the parts of your program that work well and the parts that are problem areas. Look at your existing safety and health program activities and determine their successes or failures. Study your records (accident, injury or illness data, workers' compensation rates) to see what they tell you.

Take a Good Look Around

Look at physical conditions that currently exist (OSHA violations or other hazards). This information can help you identify your workplace problems and see what's involved in solving them.

If you come up with an excessive number of physical problems such as OSHA violations, get these fixed *before* you attempt to set objectives. Not only are you vulnerable to an OSHA inspection, you are also putting your employees at risk. Further, obvious violations of safety and health standards are also obvious to your employees. Your corrections will demonstrate your interest in their safety and health.

Determine What You Need to Get Where You Want to Be

Now that you know where you stand, what do you need to get done? This is another opportunity to get employees involved in the development of your program. Positive acceptance of safety and health can result if you allow employees to suggest and participate in setting program objectives. This helps create a sense of involvement and commitment to your safety and health program.

Objectives are statements of results or performance to be achieved. They are short-term, definitive contributions to be made to the company's goal. Workplace objectives for safety and health are similar to those you set for other business functions such as sales or productivity. They identify *what, when,* and *how much,* but do not include a justification for why they should be done (that is included in your policy statement), nor a description of how they should be accomplished (that is included in your action plan). Establishing objectives involves identifying, setting, writing, and reviewing.

Identify Your Objectives

Anything can become an objective, from setting up safety/health committees to investigating accidents or developing an orientation program for new employees. You must decide what activities are most important to your program goal, and which will help you create an effective overall safety and health program. The objectives you select should be consistent with your basic safety and health policy and be part of the normal business of your company rather than special projects added on to the normal workload.

Setting Objectives

Objectives should be based upon **performance measures,** that is, indicators that tell you whether you did or did not get the performance expected. When setting objectives, you need to keep the following points in mind:

1. Relate to some part of your overall goal.
Example: An objective to develop and implement a program to train and license fork lift truck drivers relates to the part of your goal to ensure that all employees

understand the hazards and potential hazards of their work and how to protect themselves and others.

2. Be aimed at specific areas of performance which can be measured or verified.

   Example: An objective "to improve safety and health performance next month" is too general to be useful. It would be better to say "to make weekly inspections and make certain all hazards found are corrected within 24 hours."

3. Be realistic and attainable but still represent a significant challenge.

   Example: An objective for the plant to reduce recordable injuries in the upcoming year by 100 percent may not be attainable because of the extent and complexity of the steps you may need to take to prevent all injuries. An objective that is well beyond reach can soon create a defeatist attitude on the part of all those who are working toward its achievement. On the other hand, an objective for the plant to reduce recordable injuries by only 5 percent in the next year can also create a defeatist attitude. Keep in mind that some objectives will require special efforts in order to achieve them.

4. Have input from as wide a range of employees as is practicable. Even when your ideas already strongly influence your supervisors, you will find that your safety and health objectives will be most effective when you discuss them beforehand with your supervisors and employees. It will be most helpful to secure at least their agreement or cooperation. It will help motivate them to meet the objective if they feel they helped create it or at least were consulted about it.

5. Be understood by all involved supervisors and employees. Use terms that have a clear meaning to all

those directly concerned and that leave no doubt as to what is to be accomplished.

Example: An objective "to investigate all accidents and incidents to determine their cause(s)" may mean little to a supervisor. State it in clear terms: "Investigate all accidents and incidents at once to determine all contributing causes and take corrective action within 24 hours of completing the investigation."

6. Be consistent with obtainable resources. An objective that requires a very large outlay of money or a substantial increase in personnel during a period when that would not be possible is a waste of time and effort. This does not mean that the objective should be discarded. Instead, your immediate objective should be to work to produce the resources necessary to accomplish the longer-term objective. In the meantime, work incrementally at it until the resources are found.

Writing Objectives

Put each objective in writing. This will give it more importance and help you keep track of where you are in your effort to accomplish it. Spell out, in concrete terms, what is to be achieved, to what degree, and by when. Use very specific terms focusing on performance. You may also want to include a statement on the maximum amount of time or money you want to invest in achieving the objective.

Guidelines for Writing Objectives In general, a well-formulated objective reflects the following:

- It starts with the word To, followed by an action verb.
- It specifies a single key result to be accomplished.
- It specifies a target date for accomplishment.

- It is specific and quantitative (measurable and verifiable).
- It specifies the "what" and "when," avoiding the "why" and "how."
- It relates directly to the accountable manager's role in the organization.
- It is easily understandable by those who will be contributing to its attainment.
- It is realistic and attainable, but represents a significant challenge.
- It provides maximum payoff on the required investment in time and resources as compared with other objectives being considered.
- It is consistent with the resources available or anticipated.
- It is consistent with basic organizational policies and practices.

Examples:

- To conduct weekly inspections with emphasis on good housekeeping, proper use of protective equipment, condition of critical parts of equipment, and preventive maintenance.
- To determine the cause(s) of any accident within 24 hours.
- To eliminate any hazard(s) identified during accident investigations and weekly planned inspections within 24 hours whenever possible.
- To complete one job safety analysis each month in each department, with follow-up revision of safe work procedures and employee training by the following month.
- To hold and evaluate emergency drills for tornadoes every six months and a joint fire drill–evacuation with local emergency organizations each year.

Keep copies of the written objectives, and use them in discussions with your supervisors and employees. Be sure they understand the responsibilities assigned to them under each objective and that they will be held accountable for these responsibilities.

Reviewing Objectives

Review your objectives on a periodic basis to make sure that you are getting the desired performance and results and that these results move you toward your goal. Any program or activity in which you invest time and resources on a continual basis should prove its worth. You need to keep tabs on whether the objectives are being achieved, as well as whether the desired results are being obtained. For instance, if the objectives were achieved but the department continues to have too many accidents, too many close calls, too many unsafe acts, or no improvement in conditions, then obviously different or additional objectives are needed.

Summary

Just as your journey has a destination, so too your implementation or revision of a safety and health program has a goal. To reach that goal you first write and communicate your policy, which tells your reasons and commitment to reach that goal. You communicate this policy through explanation, through action, and through example. Next you clearly state and communicate your goal. Then you lay out your route, which is your objective. To lay out a route, you first need to know where you are; that is, you must assess the current state of your safety and health program and of your workplace. Then you are ready to set your objectives. Each objective must be attainable and yet a challenge. All people on this journey to occupational safety and health in your workplace should be involved in the planning and implementation of it. Whatever your policy says, make sure it is an honest reflection of what you believe and is appropriate for your workplace. Once you have established a policy that you believe in, make sure that it is clearly communicated throughout your organization. A policy statement that stays in a file cabinet is useless. Put it into action. Use it as a basis for setting your goal for the safety and health program and the objectives to meet that goal. Make sure it permeates your management systems and planning, hazard identification, hazard prevention/control, and training activities. Talk about your policy often, and let your actions reflect its content. With your policy set and communicated, you have now determined the reason for your journey. The planning of the rest of your journey becomes simpler.

Setting your goal for your safety and health program is similar to choosing a destination for a journey; it requires a determination of where you want to be. Finding the best statement of that goal is not easy and requires careful thought. Your goal could be either numerical or descriptive, but there are difficulties associated with each type. Remember, you will set your objectives on the basis of the goal and you will be evaluating your results against both the objectives and the goal. When in doubt about what your goal should be, OSHA recommends that you set as your goal "a comprehensive program to assess all hazards and known potential hazards of your worksite and to prevent or control those hazards."

Setting objectives and measuring your results are fundamental for achieving your safety and health program goal. By establishing specific objectives and getting employees involved in meeting these objectives, you set the climate for success.

When you take the time to explain to your employees

the journey you are embarking on, they can readily contribute their own judgment and common sense to do the job well. Employees will work to achieve objectives if they know what the objectives are, understand them, and are held accountable for achieving them.

Remember these important points about objectives:

1. An objective must be measurable and verifiable, must relate directly to the company's goal, and must be consistent with your basic safety and health policy.
2. An objective must be realistic and attainable while still representing a challenge to the employee responsible for it.
3. An objective must be readily understood, in terms of wording and accountability, by those who implement it.
4. An objective's end result, when achieved, must justify the expenditure of time and resources required to achieve it.

Review your objectives periodically to be sure they are leading you efficiently to your goal. Finally, be sure your goal achieves what your policy says you want. Then your journey will be complete.

3 *Visible Top Management Leadership: Showing Your Management Commitment*

Without top management commitment to protection of workers, good protection from occupational hazards will not happen. Few would disagree with this statement, but how can you tell when management *is* committed? The Safety and Health Management Guidelines list a series of management techniques to be performed as part of a basic element called "management leadership." Of these, the clearest statement of what employees at a worksite *need to see* is the action calling for "visible top management leadership in implementing the [safety and health] program."

In this chapter, we will describe ways to provide that visible leadership for the *whole worksite.* Ideally, this means involvement in a program that demonstrates concern for every aspect of safety and health. To that end, we have included a description of a system of ensuring that contract workers are both protected and prevented from endangering employees of the owner company.

A variety of methods are used by successful top managers to provide visible involvement in the safety and health protection of their workers. Many use combinations of these methods. You should look for ideas that fit you and your worksite when you decide to improve your visible involvement. These methods include:

- Getting Out Where You Can Be Seen
- Being Accessible
- Being an Example
- Taking Charge

Getting Out Where You Can Be Seen

In recent years, the phrase "management by walking around" has been heard a lot. This describes a manager who frequents all parts of operations, getting to know the people who make it work and seeing firsthand what is working well and what is a problem. The other side of this is that when employees see the manager "walking around," they know he or she cares about what they are doing and how well they are doing it. When they see that certain areas interest "the big boss," *they* become more aware of those areas.

For worker safety and health, this style of management can be demonstrated either informally or formally. Informally, the top manager takes time while he or she is out in the plant to notice hazards and to get them corrected right away. A more formal method is to conduct inspections.

Informal Action

A manager who stops to get hazardous conditions or practices corrected as he or she walks through operations areas will impress workers with the importance of health and safety. The manager should be particularly aware of short cuts in safe work procedures that are being taken to speed production. The involved manager knows that short cuts that leave out safety and health precautions are another form of Russian roulette. It is only a matter of time until an employee—and the company—get hurt. No worker or supervisor wants to have the top manager stop the work until it can be done correctly. They will make stronger efforts to do it right the first time.

If you also stop occasionally to comment on *how well* the work procedures are being followed, you will have a strong positive reinforcement impact on that desired behavior.

This informal activity, however, should be a fairly routine occurrence. If it only happens "once in a blue moon," it will not have much impact. And it only works for managers who *are* out in operational areas several times a week, if not several times every day.

Catching hazards and getting them corrected also requires a thorough knowledge of what is safe and healthful. A top manager who is not sure of his or her knowledge should not try to interfere with lower-level managers and supervisors who do know what is safe.

This informal style is well suited for the small business where the owner-manager is, of necessity, frequently in the operations areas.

Formal Inspections

A more formal method of getting out where you can be seen is to conduct surprise inspections periodically. This must be done with enough frequency that it makes a difference. The most usual form of inspection done by top managers is for housekeeping. The plant or site manager does not have to be a safety or health expert to be able to spot housekeeping violations or problems. Good housekeeping contributes to good safety and health and, to a lesser extent in most workplaces, to healthful conditions. Some managers give positive or negative points on these inspections and award prizes or a rotating trophy to the department that does the best.

Another way to accomplish much the same result is for the plant or site manager to accompany the regular safety committee or safety and health professional inspections on a surprise basis. Once again, the frequency of these surprise activities is important.

Being Accessible

If you can manage "by walking around," finding the opportunity for employees to ask questions and point things out to you is fairly simple. Even if your managerial work precludes spending a lot of time in the operations areas of your worksite, you can provide employees the same kind of opportunity through more formal systems. You must be careful, however, in any of these accessibility systems, not to undercut the authority of the managers and supervisors to whom you have given the primary responsibility for ensuring safe and healthful work and working conditions. Being accessible for health and safety concerns means walking a careful line between discouraging your employees from using that access and interfering with their normal relationships and responsibilities.

Informal "Instant" Access

Once again, this informal method is well suited to the small business owner-manager. If you *do* get out into operations areas frequently, encourage your employees to talk to you about any problems they perceive in getting the work done in a safe and healthful manner. (This helps in areas other than safety and health as well.) If you take their concerns and questions seriously and make sure that they get timely and appropriate responses, they will continue to make sure you know what is troubling them.

The Open Door Policy

If you find that your managerial work keeps you in your office, you might institute an "open door" policy, which requires that your office door actually remain open, at least during specified hours or days that do not change and are well communicated. (This probably will not work for managers who must have frequent closed door meetings.) Employees should be encouraged to drop by and tell you about safety and health concerns or ask questions about things that trouble them, without fear of reprisal, *if* they have not been able to get the answers using normal supervisory channels.

Employees should *not* be required to make an appointment, since that will discourage all but the most determined. Remember, you want to make this a casual, informal tool so that everyone will feel comfortable with it. Chances are, once employees test your policy and the word gets around that your door really *is* open, most employees will not feel the need to use it as long as your other systems are working well.

The "By-pass" Meeting

If you cannot spend a lot of time in operations areas and your need for private meetings preclude an "open door," you can plan to hold periodic "by-pass" meetings. These meetings get their name from the fact that they allow the top manager and hourly employees to "by-pass" the structure in between and talk directly together. These meetings are usually open for *any* questions, comments, or concerns that employees might have, but they are particularly useful for those involving health and safety. The size of this group should probably not exceed more than two hundred people, so larger worksites will need to divide employees into smaller groups and have several meetings. (Some managers also meet with first-line supervisors separately and other groups of lower-level management with whom they do not meet in the course of regular business.) You may need to experiment a little with the size of the group to get one that fits your style best.

The top manager should avoid preparing more than an encouraging opening or welcoming statement. This meeting should not be used to pass planned information downward. The employees might, however, need some encouragement to get started. If you decide to start having this type of meeting, you may wish to have your supervisors prepare the way during their own meetings with employees, helping them think about what topics to raise.

The success of this type of meeting will depend on the way you as the top official at the worksite handle the questions and concerns that are raised. They should all be treated with respect even if, from your perspective, the answer seems simple or the concern unnecessary. Take a few seconds to imagine how the situation looks to the employee and take the time to give a clear explanation. When you are asked questions to which you do not know the answer, or you are presented with concerns where you need to know more about the circumstances, do not be afraid to say so. Be sure, however, that you do follow up thoroughly and that all the employees who attended the meeting see or hear your response.

The Birthday Lunch

This is another, more personal version of the by-pass meeting. The plant or site manager provides a periodic lunch with all the employees who have had a birthday in a certain time period. The size of this group should be kept small (approximately 20). You may need to experiment to find out what size group works best for you. The date of birthdays will provide a "random" selection of people from all parts of the worksite.

In a low-key manner, the manager should turn the conversation to the questions or concerns employees might have. In a smaller group such as this, speaking up about perceived problems may be more frightening. You should aim for a warm atmosphere to encourage a frank exchange. Otherwise, most of the techniques for the by-pass meeting also hold true of the birthday lunch meeting.

Being an Example

One of the most important aspects of visible management involvement is providing a good example to your employees.

Following the Rules

Make sure you know all the rules that employees must follow and that you and your subordinate managers follow them scrupulously. Some of the rules may only pertain to people who will be in the area for several hours or who will be working specifically with special equipment. To the extent practical, you and your managers should follow those rules as well, even if you are only visiting for a few minutes and will not be working directly with the equipment.

Setting an Example for Supervisors

Never let an infraction of the rules or safe work practices go unnoticed if you see it. You may not always know every precise work practice required to do a job, but if you *are* sure and you see an infraction, make sure it gets corrected. If you are not sure, check immediately with the supervisor or area manager. Your insistence on working in a safe and healthful manner will set a good example for your supervisors.

Taking Charge

Make it clear to everyone working at your worksite that *you* are in charge of ensuring that it is a safe and healthful place to work. One visible way that is widely used in the chemical industry is for the site manager to chair the central safety committee. But taking charge of safety and health protection also requires that you hold your subordinate managers and supervisors clearly accountable. It further requires your leadership in providing that any contract work done at your site be done in the same safe and healthful manner that you insist upon for your own employees.

Chairing the Central Safety and Health Committee

In its usual form, the central committee is made up of the worksite executive staff. At some worksites, membership also includes two or three positions for hourly employees. Employee membership is sometimes rotated through the hourly workforce to provide a training and awareness experience. This committee should not be confused with a joint employee-management committee.

By chairing this committee, you are demonstrating to your subordinate managers and employees that you are taking charge of safety and health protection. For this to succeed, however, the committee must have serious tasks to accomplish and meet at least monthly. To be effective, you must take *your* attendance and participation seriously as well.

Accountability

No matter what systems have been set up to provide for formal accountability, your employees at all levels will feel the importance of those things to which you personally pay attention. If you never inquire about your subordinate managers' or supervisors' safety and health duties, they will quickly assume that you do not care. It is particularly important that you insist that they carry out their responsibilities to ensure that subordinate supervisors require employees to follow safe work practices.

Ensuring Safe and Healthful Contract Work

The actions of contract workers at your site can have an adverse impact on the safety and health of your employees as well as themselves. Where contract workers and your own employees are intermingled, unsafe practices or conditions of contract work will expose your own employees to hazards. But even if contract workers are constructing a building somewhat apart from your normal operations, your own workers will have their awareness strengthened by the knowledge that you insist on good safety and health practices and protection for *every* worker at your worksite.

Bidding Process

Even before you request bids on contract work, you should make clear—as prebid qualifications—your insistence that contract companies meet several requirements. You should specify an acceptable level of experience modifier rate (EMR) as set by the contract company's insurance company. You should require an implemented safety and health program.

If you have the opportunity for contact before the bids are final, you or your agent should make clear that the cost they bid for your project should include any costs for meeting OSHA standards and the rules of your worksite. Make clear to your bidders that you intend that these precautions be *fully* met.

Beware of a company that has no known experience. It may have been called something else last year. You can expect that there is a good reason for a change in name and that it probably does not bode well in terms of the performance you may expect.

Contract Language

The contract you use should spell out precisely what you expect of the company's safety and health program management. If the work it will be doing involves potential hazards to your workers and/or the community, the skill, education, and experience requirements for its workers should be specific. If you expect the company to go beyond OSHA standards in certain areas, such as fall protection on a construction contract, state that specifically. The following requirements are frequently used:

- Employee safety and health orientation and periodic safety and health training/meetings
- A formally established relationship with a physician, and contractor employees at the site who are trained in first aid
- Regular safety inspections and, where applicable, industrial hygiene monitoring
- An appropriately trained safety and health coordinator

There should also be specific language in the contract giving your agent the right to:

- Monitor safety and health activities
- Investigate contractor accidents/incidents
- Require that any worker who continues to violate safe work practices be removed from the site
- Remove the contract company from the site if the requirements of the contract are not being met

Further, the contract should clearly require that your agent will be *informed* of all chemicals or other hazardous substances to be brought onto the worksite *before* they arrive.

Contract Work Monitoring

The areas where contract work is being performed should be included in your routine general inspections. Unsafe work or work in violation of any part of the contract should be stopped and corrected through the appropriate supervisor, if possible. Your agent should check to make sure that the contract employees are informed not only about any chemical hazards to which working at your site may potentially expose them, but also about any hazards to which their own company's work may expose them. Obviously, your employees will need to know about and be prepared to protect themselves against any contractor hazards to which they are potentially exposed.

Arrangements should be made to include the contract employees in evacuation and other emergency drills. Plans should also be made for handling vendor employees and visitors to the site.

Follow Through

Use the safeguards that you put into the contract. If you discover that after several warnings, inexperienced laborers are still being assigned to work involving significant hazards in spite of the contract clause requiring training, cancel the work and reopen bids. It may cost some time but that cost is insignificant compared to the potential loss of time, money, and/or lives if the wrong moves are made with some high-hazard materials.

If your agent is finding that certain workers are continuing to violate safe work practices after the workers have been corrected, have them removed from the site. If the contract company itself is violating rules or refusing to make corrections, close the contract.

Let it be known throughout your community that at your workplace, *only safe and healthful work is acceptable.* You will find in the long run that contract companies willing to insist upon safe and healthful work will also be the most efficient and cost-effective.

Example of How the Process Works

As the owner or top manager of a worksite, you have delegated responsibility to your subordinate managers and supervisors. You do not want to undercut their authority to get those responsibilities accomplished. Yet you want to show your commitment, involvement, leadership, and *accessibility* to employees at all levels in protecting your workers from safety and health hazards. How do you do this?

1. By putting a *complete* safety and health program in place
2. By firmly holding your subordinate managers and supervisors accountable
3. By encouraging employees to use the routine systems afforded to them by the safety and health program
4. By forging a "partnership" with subordinate managers and supervisors for encouraging employees to speak out and to use the systems

Complete Safety and Health Program

Each one of the methods you use to improve the quality of your safety and health protection will work better if it is complemented by the other methods of good management. A good safety and health program cannot be pulled apart to allow a manager to use some of it without the rest. For you to show visible involvement, commitment, and leadership, the place to start is with the implementation of a whole program.

By providing a complete safety and health program, you will have given both your managers and supervisors and your lower-level employees the tools to work together with you to keep the worksite safe and healthful. This type of complete program will avoid pitting one level of employees against another.

Subordinate Accountability

If you are holding your subordinate managers and supervisors accountable for their safety and health responsibilities, the chance will be greatly lessened that they will be the *blockers* that keep safety and health problems from being resolved or keep better ideas for hazard prevention and control from being considered. If subordinate managers and supervisors are not considered part of the problem, then the question of blaming them or undercutting their authority should not arise.

Encouraging the Use of Program Systems

If you have a full program in place and you are holding your subordinate managers and supervisors accountable for carrying out their parts of it, the next step is to encourage your hourly and/or lower-level employees to use the systems in the program and to carry out their responsibilities.

You should encourage employees to make full use of the systems you have put into place for their participation in problem identification and resolution and in the system for them to report hazards. Then make sure that they get appropriate and timely *management* response when they do.

If these systems are working well, your employees' need to let you know about problems and concerns will be limited to the big problems that the normal systems cannot deal with. Your supervisors will understand that these problems are not a reflection upon them and that you are the proper person to whom to address these concerns.

An employee who brings a problem or suggestion to your attention should be tactfully asked (after the problem or suggestion is explained) what attempts were made to resolve the problem, or to have the suggestion considered, through the systems that are in place. If the employee tells you that he or she has attempted to tell the supervisor but has gotten no action, that indicates a problem within the system. The problem may be as simple as a supervisor who does not want to have a safe or healthful workplace, but that is unlikely. You should deal with the problem as a *system* problem rather than as a personal issue with your subordinate managers and supervisors. The supervisor in question may not have understood the issue raised by the employee or may not have been able to articulate to the employee why no action was necessary. Make clear to the supervisor's manager that the problem within the system should be resolved. If at least part of the problem is the attitude of the supervisor, give the manager and supervisor the chance to work it out. It is not a good idea to call the supervisor on the carpet simply on the basis of this one problem.

Obviously, if your accountability system is going to work, any individual who continues to be a "blocker" in your systems will have to be held accountable for his or her inability to do the job assigned. The important point is to try to keep that accountability activity distant from your immediate response to employee-raised questions, concerns, or suggestions.

Remember, too, that part of your systems for employees to have input into problem identification and resolution is the protection of those employees from discriminatory actions, including unofficial harassment.

Forging a Partnership

Make sure your supervisors understand that you are aware that not every safety and/or health problem can be resolved at the supervisory level. Call upon your subordinate managers and supervisors to help make the systems for employee input work. Consider them as the leaders of groups that work together to get all problems identified and resolved through whatever mechanism in your systems is needed.

You may wish to recognize and/or reward the groups that do this most successfully in terms of numbers and *quality* of reports of hazard or suggestions for better hazard control. Let your subordinate managers and supervisors know that you consider the fact that an employee brings something to your attention to be a good reflection upon his or her supervisor's leadership.

The key point to remember is to make the use of your accessibility and of all of the systems of your safety and health program a positive experience. It is part of everyone's job to find problems and get them corrected before they cause harm. If this concept is clearly communicated, you will walk the "fine line" in championship style.

Summary

As the owner or top manager at the worksite, you can make a major difference in the quality of safety and health protection of your workers. There are many different ways and styles for achieving this impact. Small businesses are probably better suited for the most informal of these methods and styles. The owner-manager should be seen as someone who is vitally interested in worker safety and health, sets a good example, and is clearly in charge of a worksite where safety and health outweigh all other considerations.

4 Employee Involvement

The men and women who work for you are among the most valuable assets you have. Their safety, health, and good will are essential to the success of your business. Your job is to develop a safety and health awareness that surrounds every employee on his or her job. You are much more likely to be successful in that effort if you involve your employees in every part of your safety and health program.

While it is true that you, as the owner-manager of the worksite, have the responsibility for ensuring employee safety and health, getting your workers truly involved in your site safety and health program is the key to making it work well. When your workers are involved in safety and health, you get the benefit of their special knowledge and interest to develop and maintain programs to protect themselves and their fellow workers.

Providing for and encouraging effective employee involvement is a key demonstration of management commitment and leadership. Your employees need to know that you want them to help you make the safety and health program work. Their involvement must be meaningful so that they will have a sense of responsibility about the success of the overall program. *The only limits to what employees can do well are those set for them or those resulting from inadequate training.*

This chapter explains some reasons for getting employees involved, ways that they can be involved, the differences between union and nonunion worksites, and what management must do to make it work. It also provides some examples and some advice on getting started.

Why Should Employees Be Involved?

Employees should be involved because it is both the right and the smart thing to do. These are the reasons why:

1. Line workers are the persons most in contact with potential safety and health hazards. They have a vested interest in creating programs to protect themselves against these hazards.
2. Recent experience has demonstrated that line workers make highly valuable problem solvers.
3. Group decisions are usually better than individual ones because they bring a wider field of experience to bear.
4. Research shows that employees are also more likely to support and use programs in which they have had input.
5. When employees have the opportunity to put forward their ideas for serious consideration, they enjoy their work more and become more productive.

Close Contact with Hazards

You as the owner-manager of your worksite probably know more about the overall operations there. Line workers, on the other hand, have the most detailed knowledge about each operation of your worksite. They are the ones who actually perform the tasks that will either expose them to or protect them from the potential hazards in the work.

If they have a good understanding of the hazards, they will also understand that they have the most to gain from preventing or controlling exposure to themselves and their fellow workers.This understanding will also provide you with a good source of ideas for better prevention and/or control of those hazards.

Value as Problem Solvers

The Japanese have used their workers for many years to help solve various kinds of workplace problems. American companies do not have as much experience yet, but where this concept has been tried, it has met with success. Currently, worker input in the United States is mostly used for quality control problems. This has been successful for the companies using it. Safety and health protection problems are even better suited to worker involvement for the reasons cited above.

Improved Support

Frequently managers complain that they just cannot get workers to do what they are supposed to do to work safely, whether it is wearing the appropriate personal protective equipment or following safe work procedures. How do you change that? It is simply part of human nature to support the ideas that we help develop and put to work better than those that others force upon us. Line workers are no different. If they have a chance to develop or revise rules and/or safe work practices, they will have a personal investment in making sure that those rules or procedures are followed. If you involve employees in ruling making and enforcement is still a problem, invoking the disciplinary system is still an option.

Value of Group Decisions

Decisions by committee have a bad reputation, because committees may not always be the fastest or most efficient method of reaching a decision. But we now know that group decisions can, in many cases, be the best. If you try to have a group reach a decision with which all its members are comfortable, you must allow all points of view and realms of experience to be considered. Improved information means improved decisions.

There are several group exercises that demonstrate this by asking first individuals, and then groups of those individuals, to rank items in terms of their usefulness for survival purposes in different hostile circumstances. Both sets of ranking decisions are then compared against decisions made by experts. In these exercises, it is rare for individuals to make better decisions in the ranking of those items than the groups do, even though the groups are made up of the *same* individuals.

More Involvement Means Better Work

When your employees have the opportunity to help you find problems and then develop the solutions for the problems, your employees enjoy their work more than if they simply do what they are told to do. When workers enjoy work (and we mean work, not pleasurable ways to avoid work), they work faster and produce a better-quality product. Since they are also less likely to look for jobs elsewhere, turnover is therefore reduced.

What Can Employees Do To Help?

Employees can be involved in just about any safety- and health-related activity in which you might want them to participate. OSHA has seen employees involved in a variety of activities, including but not limited to:

- participating on joint labor-management committees and other advisory or specific purpose committees
- conducting site inspections
- analyzing routine hazards in each step of a job and/or a process and preparing safe work practices or controls to eliminate or reduce exposure
- developing and revising the site safety and health rules
- training of both current and newly hired employees
- providing programs and presentations at safety and health meetings
- conducting accident/incident investigations
- all-pervasive employee participation

Some of these activities require training before employees can be proficient. The training is not elaborate and can be provided at your worksite.

Committee Participation

Joint labor-management committees are the classic method of employee participation. They are extensively and successfully used in many European countries and Canadian provinces. Other types of committees have also been used successfully for safety and health participation. Many union worksites have employee safety committees where the employee representatives—selected by the union(s) or elected by employees—work alone without management on varied functions. At some worksites, there are hourly workers participating on the central safety committee, which is the executive staff of the worksite. In addition, some worksites use employee or joint committees to serve specific functions, such as inspecting the site for hazards, investigating accidents and incidents, and training new employees. Finally, although they are not called "committees," quality circles are another form of committee that focus, at least part of the time, on health and safety problem identification and resolution.

Classic Joint Labor-Management Committees

These committees usually have equal representation between labor and management. Chairpersonship may rotate between an employee representative and a management representative. The "powers" of the committee are worked out through negotiation in great detail. There are usually quorum requirements, and votes are formal. Although the functions of these committees do vary depending upon the outcome of negotiations, the most usual committee functions are site inspections with oversight of hazard corrections, investigations of employee reports of hazards, accident investigations, and safety and health awareness program development. Sometimes the committees simply receive reports from the experts on these activities and monitor hazard correction and program effectiveness.

Other Joint Committees

Other joint committees are found with either more employee participants (construction sites with many different trade unions to be represented) or more management participants (if medical, safety, and industrial hygiene personnel are counted as "management"). They are frequently chaired by the highest ranking safety "specialist" at the site but are sometimes chaired by an hourly employee elected by the committee itself. They work by consensus and do not have formal votes. Their usual functions are similar to the classic joint committees.

Employee Safety Committees

These are usually union safety committees. Their membership is determined by the union. Some worksites with more than one union will have more than one union safety committee. The committee operates without direct management staff involvement but has regular meetings with management and management staff. At these meetings, the committee's concerns are raised and responses received. The committee may do inspections and investigate employee reports of hazards but will usually carry their findings to management for action. They may also design and oversee employee awareness programs.

Central Safety Committee

At nonunion sites, particularly in the chemical industry, central safety committees are made up of the site manager and the executive staff. In recent years, some companies have discovered that it is helpful to have hourly worker representation on this committee. Some sites rotate the employee positions on the committee throughout the population of the worksite. At other sites, the hourly workers are selected by management for their experience and achievements in other safety and health employee participation systems.

The central safety committee is an oversight committee with an interest in every part of the safety and health program. It sometimes serves as the hazard correction tracking system, with the committee receiving reports of all inspections, accident and incident investigations, and employee reports of hazards.

Specific Function Committees

Some companies have used "permanent" single-function committees very effectively. They usually allow employees to volunteer on the basis of the function statement for the committee. These committees can comprise only employees, with management liaison; or they may be joint committees with some management and/or safety and health staff (including plant nurse or doctor). Each committee has a single function, such as accident/incident investigation, site inspections, site safety and health rules, safety and health training, or safety and health awareness programs. The committee members are trained and provided with the resources they need, such as site safety and health experts to assist them; reference materials; and film, videos, and/or equipment, such as cameras, for accident/incident investigations or inspections.

Quality Circles

Quality circles are work groups that are generally formed to address quality problems. They spend some of their time together brainstorming problems and ideas for solutions. They frequently address problems involving safety and health protection. This form of committee may work best for smaller worksites where there are really not enough employees to put together other types of committees. The circles can address all aspects of a problem, not just quality or safety and health. They can also look at productivity implications. Just make sure that they are also helping you find and resolve safety and health problems.

Conducting Site Inspections

One of the most frequent areas of participation is the site inspection. This can be done through a joint committee or an employee committee as one of several functions, or it can be a single-function committee. Another method of using employees to inspect is the safety observer type of inspection. Whatever the method, these employees *must be trained* in how to recognize hazards that are possible in your industry. They should also have access to your safety and health "expert(s)" and to written references. For true participation, the committee or safety observer should be able to suggest methods of correcting the hazards that they find and track the correction to completion.

Committee Inspections

The group making the inspection should probably not exceed four people in the same area at a time. For larger committees with several functions, inspections can be done by a subcommittee. If inspections are the only function of a large committee, the inspection duties can be rotated or groups can be assigned to different parts of the worksite.

Safety Observers

Some workplaces have safety observers who check their work areas for hazards periodically. Some do checks every day for the first few minutes of the shift. Others do more thorough inspections weekly or monthly. The frequency will vary with the nature of the hazards and the size of the worksite. Safety observers usually work with the area supervisor to get hazards corrected. They also usually work alone, although some companies have their safety observers come together as a committee monthly or semiannually to brainstorm problems or ideas for improvements that go beyond their own work areas. It is important to remember that the safety observers must also be involved in the correction of hazards for the participation aspect of their activity to be effective.

Routine Hazard Analysis

Employees can be very helpful in taking apart jobs, processes, or activities that arrive in phases, to examine the hazards that may be hidden and to design improved controls through revised work practices or processes. These activities are frequently accomplished by teams that include supervisors. For complicated processes, the teams will probably be led by engineers. Many companies find, however, that workers who use the procedures or processes every day make excellent analyzers for this work.

Revised practices or processes that result from these analyses will be better accepted if workers are involved in the decisions that are made.

Developing or Revising Site Safety and Health Rules

Asking a group of your employees to be responsible for developing or revising and updating site safety and health rules can be very rewarding. If employees help make the rules, they are more likely to obey them and remind others to obey them as well. Their input may make the rules better or stronger because of their special knowledge, not only of the nature of the work, but because of the nature of their fellow workers. In a large worksite, this could be a standing committee. At smaller worksites, it could be a periodic group meeting or part of a more general employee or joint committee set of functions.

Training Other Employees

Using your best or longer-term employees to teach newly arrived employees can be very effective. It may also breathe new life into your ongoing periodic safety and health training to have employees take responsibility for it. Many companies use one or both of these means of employee participation to good result.

New Employee Orientation

Your hourly employees can make excellent instructors for new employees. You will want someone in management to go over the personnel–employee relations portions of the orientation, but any part of the rest of it can be done by hourly workers with appropriate training. In particular, the part of the orientation that goes through the new employee's job can be done very effectively by other, experienced workers whom you have trained to perform this role. The trainer can follow up the initial introduction to the job by being a "buddy" to watch over the new employee, giving advice and answering questions that a new employee might be afraid to ask of a supervisor.

Ongoing Periodic Training

Many companies have found that making employees responsible for regular safety and health sessions has been good for keeping interest in the training high while at the same time improving overall safety and health awareness. Doing this requires some help from management in getting started and in providing ongoing assistance with new training ideas, materials resources, and references. This can be done through a special committee, as part of the more general functions of an employee or joint committee, or just by rotating employees through the work group without a committee designation.

Program Presentations

Program presentations for safety meetings and other training are discussed above. In addition, employees can be used to plan and present awareness programs that include award programs, poster and slogan contests, safety and health newsletters, and the like. If you decide to include some sort of award or reward program, it is very important to remember *never* to encourage employees to underreport injuries or illnesses by establishing rewards for "hours worked without injury" or similar concepts.

Accident/Incident Investigations

A frequent participation function for employees is assistance in conducting accident/incident investigations. This can be done as one of the functions of a committee with several functions or as the task of a special single-function committee. For this to be successful, the employees doing investigations need special training and appropriate equipment to do a good job.

All-Pervasive Participation

Although still a small fraction of American workplaces, the number of facilities where employees are involved in all aspects of decision making is rapidly growing. Only a few years ago, this type of participation appeared to be limited to worksites with highly trained and specialized employees. Now workers with only on-the-job training can be found doing statistical analyses of their production lines with computers, checking on product quality and production efficiency, and looking into safety problems. Operator groups work closely with engineers in designing changes to improve quality, efficiency, and safety. This degree of participation really does not require a separate program to involve employees, since participation is already built into all processes.

For companies that have not yet tried employee participation, this form is probably not the best way to start or invigorate less successful participation systems, unless the workplace is very small or is composed of highly educated, trained, or skilled employees with a lot of management interaction. As a long-term goal, it may be desirable for any type of industry or workplace.

Different Approaches: Union versus Nonunion Sites

Successful employee involvement at unionized worksites is achieved differently from that at nonunion worksites. Neither type of worksite, however, is less conducive than the other to success in employee participation in safety and health programs.

Unionized Sites

Characteristics

Cynthia Burton and Edward Cohen-Rosenthal, quality of work-life experts, describe employee participation at unionized workplaces as "building a mutually sustaining and rewarding marriage between two equal partners." They say that a union is a political organization with its own accountability system. As an organization, it will also have its own culture, rituals, and practices. You must accept and blend these special characteristics into the agreed-upon approach for a cooperative effort to succeed. The union will need to be involved from the very beginning and help work out all aspects of this joint project.

Since reducing occupational illness and injury is clearly a benefit to both workers and management, improved safety and health is usually a good candidate for a mutual goal for union and management. The goal is usually easier to reach, however, if safety and health is removed from the usual collective bargaining channels. For example, some worksites have a clause in the collective bargaining agreement that safety and health is not subject to negotiation. Others additionally require that the personnel involved in the safety and health cooperative effort will be different than those involved in contract negotiations or grievance resolution.

Usual forms

The form of cooperative, participative effort is usually a joint labor-management safety committee. Sometimes, however, an all-employee safety committee will be the method used. (See discussion of committees above.)

Responsibilities of these committees can range from reports of hazards with correction suggestions through site inspections to accident investigations. Some committees are advisory; others have specified powers to achieve correction of hazards and, sometimes, to shut down unsafe operations.

Often joint labor-management committees are given inspection responsibilities and advisory activities concerning employee complaints or accident investigations. Hazard correction is usually accomplished through the powers of the safety department as part of the committee work.

Nonunion Sites

Characteristics

If you are accustomed to union workplaces, you may feel that it is impossible to have true employee involvement at nonunion sites. If you have always worked in nonunion workplaces, you may feel that attempting to set up employee involvement will result in unionization. Neither view is necessarily correct. As Burton and Cohen-Rosenthal point out, "In nonunion places, the methods require building one big happy family."

If you are going to start up a system of employee involvement, you may have to overcome a lot of hesitation on the part of your workers. Since they are not backed by the power of a union, they may need considerable encouragement to realize that their participation is desired and is taken seriously. You will have to start building a whole company or worksite "culture" of the "family." Since you do not want your employees to think of management as "them," you will need to stop thinking of workers as "them." All of you must begin to think about "we" and "us."

At both union and nonunion worksites, employee involvement relies on *respect*. At union sites, it is between *representatives of organizations;* at nonunion sites, it must be between *individuals.*

It is a good idea to start by asking your employees at all levels for ideas for getting all of them involved in problem identification and resolution. That can serve as the first participative effort.

Usual forms

Generally speaking, there is wider variety of employee involvement opportunities at nonunion sites. Each one of the forms discussed above under the section entitled "What Can Employees Do to Help?" has been used at one nonunion site or another.

Selecting Employees

Because nonunion employees are not acting as representatives through an authorized bargaining agent, they are really representing only themselves. This means a completely different approach to participation.

At nonunion sites, employee involvement is frequently rotated through the whole population of workers for the widest possible dispersal of experience. The benefits of this approach are in educating and increasing safety and health awareness of the whole "population" of workers.

At other nonunion sites, employee involvement relies on volunteers only. Still others involve employees who are appointed by their supervisors. In both these cases, there is an improved chance for accumulation of expertise.

The best method for your worksite will depend on what you want to achieve with employee participation. If improved employee awareness is a major objective, rotational programs are a good choice. If high-level skill and knowledge are needed for safety and health objectives, volunteers or appointees that hold positions for several years would be better.

What Management Must Do

As the owner-manager, you set the tone. Unless you are totally behind this move to get employees involved, it will not work. There are two sure "killers" of employee involvement programs: lack of management sincerity in asking for employee involvement, and lack of employee belief in management sincerity.

Managers sometimes claim that safety committees, for example, may only want to talk about "trivial" things like cafeteria menus. They may decide from that evidence that employees are either unwilling or unable to address the hazards at the worksite.

Employees, on the other hand, often do not believe management actually wants their input on serious matters, so they may fall back on something "safe" such as cafeteria menus. If not corrected through commitment and positive action, the resulting mistrust and miscommunication between management and employees will defeat the intended purpose.

Here are some things that you must do to make employee involvement work in any type of workplace:

1. Be convinced that you will have a safe and healthful workplace, whatever it takes.
2. Show your commitment through leadership.
3. Make clear to your employees that a safe and healthful workplace is a condition of their employment.
4. Get as many employees involved as possible, brainstorming, inspecting, detecting, and correcting.
5. Put their safety and health participation work "on the clock."
6. Make clear to your employees what you expect of them.
7. Give them adequate training and resources for the job expected of them.
8. Take their work products seriously, implementing their safety and health suggestions in a timely manner or taking time to explain why it cannot be implemented wholly or in part.
9. When their ideas are successful, make sure that their fellow employees hear about it.

Your Conviction

If you are determined to have the safest, most healthful workplace feasible, you will insist on a complete safety and health program.

Commitment Through Leadership

When you pitch in and get involved in the safety and health program, providing leadership and setting the example, employees will know that you are serious about wanting to make their workplace hazard free.

Safety and Health as a Condition of Employment

Give everyone at the worksite some responsibilities for safety and health and hold them accountable for meeting those responsibilities. Make sure that everyone understands the rules and work practices that are designed to keep work and the place of work safe and healthful. Use positive reinforcement and, when necessary, fair and consistent discipline for enforcement.

Getting Many Employees Involved

As mentioned above, everyone should have some responsibility for safety and health; the more the better. Ideally you will also use several methods of employee involvement to tap into the greatest reservoir of ideas and abilities.

On the Clock

Since employee involvement in the safety and health program benefits the company bottom line, it makes no sense to ask employees to work on safety and health on their own time. Your investment in their time will pay you back several fold.

Clear Expectations

You should personally take time to tell the employees who are working on employee involvement projects that you are counting on good ideas and complete work and that you will not accept less. If they know that you have high expectations, they will try hard to fulfill them. Do not accept trivial, "cafeteria menu" complaints; that is not why you have given them work time to improve safety and health protection. Be sure that they know what you do expect.

Adequate Training and Resources

If you ask employees to make inspections, make sure they learn what to look for during those inspections. If you ask them to investigate accidents/incidents, make sure they know how to investigate and that they have the proper equipment to do it successfully. If you ask them to plan and provide training or safety and health awareness programs, make sure that they have reference and resource materials. Make your staff and any contract experts available to them as resources, as well. If you do not provide needed training and/or resources, their chance of giving you good work products is greatly reduced.

Take Employees' Work Products Seriously

Give employees' ideas and conclusions serious consideration. Be careful not to reject them out of hand because they seem are unusual or too simple. If an idea they provide seems to have some merit, give it a try, and if possible, involve them in the "test." If it does not work after you have given it a fair trial, go back to them and ask for another idea.

Give Employees Credit

When their ideas and conclusions work or help the program, make sure employees get plenty of credit. Letting their co-workers know how pleased you are with the employees who are involved in safety and health will make getting involved in safety and health much more desirable for others in your workplace.

GETTING EMPLOYEE INVOLVEMENT STARTED

Meet with Employees

1. Meet with employees in one large group (if not unwieldy) or in groups by shift or craft, depending upon the nature of your worksite.
2. Explain the safety and health policy of your worksite and the objectives that you hope to achieve.
3. Explain that you want employees to help with the safety and health program and ask for their suggestions.
4. Try to use as many of the reasonable suggestions as possible in some visible way.

Formal Means of Involvement

1. If the group(s) of employees have not suggested a formal means of establishing their involvement, form a joint committee large enough to represent different

parts of your worksite but not so large that it is unwieldy.

2. Try to have an equal number of management and nonsupervisory employees on the committee.
3. Management members should have enough clout to get things done.
4. The safety and health staff should be on the committee.
5. If your worksite has a collective bargaining agent, allow that organization to select the method of choosing nonsupervisory members.
6. If your worksite is not unionized, you may choose to get your employees' advice on how to decide which nonsupervisory employees will serve on the committee. Do not hold an election because of the danger of violating the National Labor Relations Act.

Summary

The quality of a decision improves when a wider base of input is used. Employees are uniquely equipped to provide excellent assistance to a wide variety of health and safety program aspects. At unionized sites, the culture of the union and its system of accountability will have to be taken into consideration in any joint effort you plan to initiate. In nonunion workplaces, care should be taken to ensure employee protection from harassment for safety and health duties and to foster a "one big happy family" approach to safety and health decisionmaking.

Management will get the quality it expects from employee involvement. If you provide leadership, a good example, "on the clock" time, clear expectations, effective training, and insistence on the employees' best efforts, the results will be surprisingly good. If, instead, you only pay lip service to it, the quality of the product you get in return will reflect this attitude. Your employees will produce the product they sense you want them to produce.

Getting employees involved in your safety and health program can be very rewarding, both for you and for them. They can provide you with the benefit of their intimate knowledge of the tasks that are performed and with more experience for better decisions. As a result, they will be more supportive of your safety and health program and will enjoy their work more.

Whether your worksite is union or nonunion, you can successfully involve your employees in the structure and operation of the safety and health program. If your site is union, remember to approach employee involvement with the understanding that the union must agree to be your partner in the whole effort, starting from the beginning. If your site is nonunion, remember that your employees may need extra encouragement to get started on the right foot. Consider asking their help in the design of the system and get everyone thinking "we" instead of "us and them."

5

Assigning Safety and Health Responsibilities

As a business owner or manager, you have the ultimate accountability for the safety and health of your employees. You cannot delegate this to others in your company. You can, however, expect others to share responsibility for certain elements of the safety and health program.

In order to make your safety and health program work, all employees must understand their individual roles. One of the ways to accomplish this is by assigning specific safety and health responsibilities via job descriptions.

The purposes of job descriptions in a safety and health program are to clarify individual responsibilities and authorities, and to help distribute the responsibilities among the supervisors and employees. Written job descriptions that outline safety and health responsibilities are more useful than oral assignments, for the following reasons:

1. They leave no doubt as to the responsibilities and authority for each job.

2. They result in improved communications and better coordination of effort among jobs.

3. They aid in determining whether all responsibilities are fully covered, whether new tasks should be assigned, and when responsibilities should be added.

4. They aid in developing job performance objectives and establishing performance measurements.

If you are the owner or manager of a small operation, you have not only accountability but also full responsibility for the safety and health program. This should not be a big problem because you have an everyday working knowledge of the problems and are closer to your employees than managers of larger concerns. However, as your business grows in size, the number of people who should assume job safety and health responsibility also grows. If you have the responsibility now, don't neglect to reassign it in the future if growth in your company merits it.

For managers of large organizations, written, clear assignment of responsibility for safety and health is an absolute necessity. In both large and small organizations, making sure that authority and resources are provided is also necessary to make the assignment of responsibility work as you envision.

Job descriptions contain the basic information that describes the important characteristics and responsibilities of each position. Collectively they describe the company structure and its safety and health management system as a whole.

To establish job descriptions, you need to: (1) review your existing organizational structure; (2) determine which part each position should have in carrying out the safety and health program and the levels of authority and resources necessary to do the job; (3) determine and assign the responsibilities for each position; and (4) communicate with the people involved.

Appendix 2 is a Safety and Health Program Responsibilities Worksheet, which can be used to outline written job descriptions.

Review Existing Organization

Within every business, there are people who should be involved in carrying out the safety and health program. On a sheet of paper or a form like the worksheet presented in Appendix 2, list all the positions in your business. Use a separate sheet for each position; you can group similar positions later on.

Determine the Role of Each Position

Next, you should look at each position (or group of positions) and decide what role that position has in your safety and health program and the level of authority that should be assigned to it. Write a statement of overall responsibility and authority for each position. Although managerial and supervisory positions usually carry some authority already, you may want to make changes for the purpose of your safety and health program. If so, be sure to state the scope of authority clearly by indicating supervisory relationships, the amount of money that the position can spend, and any other measures that describe what a person in that position can do without first getting further approval. Do not, at this point, attempt to describe in detail the specific responsibilities of a job. For example:

- *The owner* establishes and provides the leadership and resources for carrying out the stated company safety and health policy.

- *Managers and supervisors* maintain safe and healthful working conditions within their respective jurisdictions, whether out in the field, in the shop, or in the office.
- *Employees* exercise care in the course of their work to prevent injuries to themselves and to their fellow workers.
- *Visitors, vendors, customers, and subcontractors* comply with all safety and health regulations while on company premises.

In addition to normal duties defined in the positions above, people in the following functions may have additional general responsibilities:

- *Safety departments* be fully responsible to the owner or manager for the direction and administration of the safety and health policy.
- *Managers and engineering departments* ensure that all equipment that could affect the safety and health of employees is selected, installed, and maintained in a manner that eliminates or controls potential hazards.
- *Purchasing departments* ensure that safety and health equipment and materials are purchased in a timely manner, and that new materials, parts, and equipment are analyzed for potential hazards so that controls or preventions can be implemented, and that such new materials, parts, and equipment are obtained in accord with all applicable safety and health requirements.

Determine Specific Responsibilities

Now that you know who should be involved in your safety and health program, you need to write out explicit statements of what each person should do to help meet the goals of your program. To help you, we have attached a worksheet that lists safety and health responsibilities common to the different levels in a business. You will also find some suggested responsibilities for several different types of employees and an example of the assignment of responsibility from a major corporation. These are job descriptions. Review the various duties on the worksheet and determine which responsibilities fit into your program and organizational structure, and which should be delegated to specific positions in your company. Write these out on the first worksheet that describes each position. Do not confuse work rules and safe work practices for assigned responsibilities when you are developing responsibilities for nonsupervisory employees. A more general statement about understanding and following rules and safe work practices is more appropriate and will result in shorter job descriptions.

Assign Responsibilities

Now you are ready to assign responsibilities. You can delegate the details for carrying out your safety and health program to the same people who are responsible for plant operations and production. In this way, safety and health becomes as much a part of the overall management system as production.

Be sure that each assigned responsibility has adequate authority and resources to carry it out.

Write a Program

After you have clarified the safety and health responsibilities of each position, you may find it useful to combine all these statements into a single document. Then post it or circulate it to all employees involved. Talk about the job descriptions and responsibilities in face-to-face discussions between you and the employees who will be responsible for each aspect of the program. Also, keep a copy and periodically refer to it with your employees in performance reviews.

Putting your program into writing can take many different forms. The format generally depends on what format the company has accepted for all programs. Examples of two formats are given below, the first being a "bullet" type format and the other a narrative format.

Example 1: Bullet Format

Assignment of Safety and Health Responsibilities

President/Owner/Site Manager

- Establish a policy to hold the worksite in compliance with all applicable federal or state standards and to provide safe and healthful work and working conditions for every person at the site.
- Provide the leadership and resources to carry out the stated company safety and health policy.
- Set objectives and support safety and health personnel in their requests for necessary information, training, experts, facilities, tools, and equipment to conduct an effective program and establish a safe and health workplace and working conditions.
- Assign clear responsibility for the various aspects of the safety and health program and ensure that those with assigned responsibilities have adequate resources and authority to perform the duties assigned.

- Hold those with assigned responsibilities accountable by checking to make sure they are meeting those responsibilities, providing correction to those who do not and rewarding those who do.
- Keep in touch with employees and their safety and health activities, assist in giving direction and authority for those activities, and let employees see your visible involvement.
- Set a good example in following safety and health rules and safe work practices.
- Require all vendors, customers, subcontractors, and visitors to comply with the company safety and health policy.
- Make sure that you have a thorough understanding of the hazards and potential hazards that employees may be exposed to at the worksite and ensure that a program for prevention and control is put in place for all potential hazards.
- Provide a reliable system for employees to report conditions and situations that appear hazardous to appropriate management members and make sure that responses to them are appropriate and timely.
- Encourage employees to use the established hazard reporting system(s) and guarantee a strict nonretribution policy for those employees, supervisors, and managers who do use the system(s).
- Establish a system of self-inspections and review the results periodically to ensure proper and timely hazard correction.
- Establish a plant preventive maintenance program to ensure proper care and functioning of equipment and facilities.
- Review accident reports to keep informed of accident causes and trends.
- Provide a medical program, emergency response,

and first aid facilities that are adequate for the size and hazards of the worksite.
- Require periodic drills to ensure that each employee knows what to do in case of an emergency.
- Establish training programs to improve the competency of all employees in recognizing and understanding hazards and how to protect themselves and others.

Safety and Health Director/Coordinator
- Maintain safety and health expertise through training, reading, conferences, and use of experts from outside the worksite.
- Keep aware of and be able to interpret laws and standards dealing with employee risk reduction in your industry and with illness and injury recordkeeping.
- Act as the eyes, ears, and conscience of top management where employee safety and health is concerned.
- Develop a complete inventory of hazards and potential hazards and plan a program of prevention and control.
- Evaluate the plant preventive maintenance program's effectiveness in ensuring a safe and healthful workplace.
- Provide analysis for the design, purchase, and/or use of new equipment, facilities, materials, and the design of new processes in terms of hazard detection and subsequent plans for prevention or control.
- Provide technical assistance and support to production supervisors and employees in their safety and health activities under the safety and health program.
- Ensure that appropriate general plant safety and

health rules are developed, communicated, and understood.

- Assist in or oversee the development of a system for consistent and fair enforcement of the rules and safe work practices.
- Assist management in the provision of adequate personal protective, industrial hygiene, safety, and fire prevention equipment.
- Inspect facilities to detect hazards that have "escaped" existing preventions and controls and hazards that have not been detected by other means.
- Oversee investigation of or investigate employee reports of hazards and respond to employee safety and health suggestions.
- Assist supervisors in investigating accidents and incidents such as property damage and "near miss" cases.
- Provide technical assistance to employees in the performance of their duties under the safety and health program.
- Assist in the development and provision of safety and health training to employees at all levels to ensure complete understanding of safety and health hazards and responsibilities for self-protection and protection of others.
- Oversee, analyze, and critique periodic emergency drills to help improve worksite emergency readiness.

Plant Superintendents/Division Managers/Directors

- Provide the leadership and positive direction essential in maintaining the safety and health policy as the major priority in all operations.
- Hold all subordinate supervisors accountable for all assigned safety and health responsibilities, including making sure that all safety and health

policies, procedures, and rules are complied with by the employees under their direction and providing in-depth analysis of all occupational injuries and illnesses that occur.

- Actively participate in and support employee participation in safety and health program activities and follow up on recommendations made by any employee (or joint labor-management) group operating under the safety and health program in an appropriate and timely manner.
- Make certain that all new facilities, equipment, materials, and processes are analyzed for potential hazards before completion of design or purchase; that all potential hazards have been prevented or controlled before their introduction into the worksite; that tools and machinery are used as designed; and that all equipment is properly maintained.
- Ensure that job hazard analyses are conducted periodically for all jobs, with particular emphasis on those known to be dangerous, in order to seek out and prevent or control newly discovered hazards.
- Make sure that employees know about and are encouraged to use hazard reporting and safety and health suggestion systems, that they are protected from harassment, that their counsels are heeded, and that their ideas, when helpful and feasible, are adopted.
- Ensure that prompt corrective action is taken whenever and wherever hazards are recognized or unsafe acts are observed.
- Make sure that all hazardous tasks are covered by specific safe work procedures or rules to minimize injury.
- Provide all necessary safety and health equipment

and protective devices and make sure they are understood and used properly.

- Ensure that all injured persons, regardless of how minor the injury, receive prompt and appropriate medical treatment.
- Ensure that all accidents and incidents are promptly reported, thoroughly investigated, and properly recorded and that safe behavior award programs, if used, do not encourage underreporting of any incident that must be recorded on the log.
- Keep abreast of accident and injury trends and take proper corrective action, when needed, to reverse these trends.
- Ensure that all employees are physically qualified to perform their work.
- Make sure that all employees are trained and, when necessary, retrained to recognize and understand hazards and to follow safe work procedures for each hazardous job.
- Ensure that supervisors hold periodic safety and health meetings to review accidents/incidents and analyze their causes and to promote free discussion of hazardous work problems and possible solutions.
- Ensure that the safety "director" is used for assistance in promoting aggressive and effective safety and health programs.
- Help develop and implement emergency procedures and ensure that all employees have the opportunity to practice their emergency activities.

Supervisors

- Supervise and evaluate worker performance, considering each worker's safe and healthful behavior and work methods.
- Actively support employee participation activities

under the safety and health program and participate as needed.

- Learn and keep up the knowledge and skills required to detect safety and health violations and other hazards such as improperly functioning machinery, tools, or equipment.
- Ensure that the plant preventive maintenance program is being followed, and that repair and replacement needs found during those activities are tracked to completion.
- Investigate accidents thoroughly to determine how the procedure or condition can be "foolproofed."
- Actively discourage short cuts and consistently and fairly enforce safe work procedures and safety and health rules.
- Provide continuing on-the-job training in safe work operations and, where necessary, personal protective equipment maintenance and use.
- Be personally scrupulous in following the safe work procedures and safety and health rules that apply to the area.
- Make sure each employee knows what to do in case of an emergency.

Employee Responsibilities

- Learn and understand the rules, follow them, and avoid short cuts.
- Review and ask questions if you do not understand the safety and health educational material posted on bulletin boards or distributed to work areas.
- Be individually responsible to keep yourself, fellow employees, and equipment free from mishaps.
- Be certain that you completely understand instructions before starting work and avoid short cuts from safe work procedures.
- If any doubt exists about the safety and/or

healthfulness of doing a job, stop and get instructions from your supervisor before continuing the work.

- Offer safety and health suggestions when such suggestions would reduce risks to workers.
- Get involved in the employee participation system and support other employees in their assigned roles under the safety and health program.
- Make sure you understand exactly what your responsibilities are for each different kind of emergency situation that could arise.
- Know how and where needed medical help may be obtained.
- Report all accidents as well as unsafe conditions and acts to your supervisor or use the system set up to allow reporting elsewhere.

Example 2: Narrative Format

Assignment of Safety and Health Responsibilities

Corporate Management

Corporate Management's Role in Safety and Health To be effective, a safety and health program must begin at the top. Corporate officers must be sincerely convinced that hazard-free performance is the company's goal and they must lead their subordinates to the same goal. The performance and attitude of line management will reflect the safety and health attitude demonstrated by corporate officers.

Reasons Behind a Corporate Safety and Health Program Management is responsible for the safety and health of all employees. The primary reason behind a corporate safety and health program is to provide a means for discharging that obligation. In addition, there are other sound, practical reasons behind a safety and health program. Reduced injuries and illnesses increase the amount of time and effort management can apply to business problems and result in fewer interruptions to production and in improved morale among employees. A successful safety and health program reflects favorably on labor relations when employees recognize that management is interested in their well-being; it helps community relations when the community views the plant or company as a safe place; and, equally important, a successful safety and health program minimizes government intervention.

Corporate Management's Responsibilities Safety and health is a product of sound management. It reflects management's efficiency and expertise. A sound safety and health program requires the use of basic management practices: organization, delegation, accountability, and the use of feedback. An unsatisfactory safety or health performance usually indicates a fundamental management problem. Therefore, the first step in organizing a safety and health program is to consider the fundamentals of management.

Management must accept the responsibility for the safety and health of its employees; thus, safety and health should rank equally with business among the corporation's goals and objectives. This corporation has adopted a corporate safety and health philosophy stipulating that safety and health are to be given prime consideration in company activities.

Once a safety and health philosophy is adopted, management, in consultation with the line organization, must establish goals for safety and health performance. Examples of such goals include improving over last year's performance by a specified percentage; establishing a ratio of lost-time injuries to a measure of production; controlling a particular type of injury, such as eye injury; or eliminating or reducing personal exposures to hazardous substances or conditions.

To provide for the achievement of defined safety and health goals, management must establish a safety and health organization. It must delegate through the line organization the responsibility and corresponding authority to implement the various facets of the safety and health program. In turn, the line organization must have the skills necessary to accomplish these tasks. Management must provide the funds needed to implement the program.

Having delegated responsibility, management must hold employees accountable for safety and health performance by including safety and health as part of employees' performance evaluations and by recognizing safety and health performance in assignments, promotions, and compensation.

To obtain feedback, management must request periodic reports for the line organization. A safety and health paragraph should be included in regular status reports. Formal evaluations and statistical reports by the Corporate Safety and Health Services Department will provide additional feedback.

Corporate Management's Involvement in Safety and Health An active safety and health program involving corporate, subsidiary, and division management is just as important as programs at plants. Members of corporate management can personally contribute to this safety and health program by supporting the corporate safety and health philosophy with action and attention. Participation, inspection, and review will demonstrate interest. Subordinates must have management's support through the authorization of capital expenditures to minimize exposure to injury and illness. Safety and health must be managed as any other business objective.

Facility Managers

The Facility Manager's Role in Safety and Health

A sound, active safety and health program provides a means for determining the performance of an operation group or facility and the effectiveness of its programs. A successful program contains all the elements that represent sound management theory: definition of the organization; delegation of authority and responsibility; establishment of goals and objectives, clearly stated and mutually accepted; requirement of accountability; and initiation of control and feedback systems.

Injury and illness statistics can be used effectively as feedback: they are leading indicators of fundamental organizational problems. Unfortunately, they alone are not reliable tools for controlling a safety and health program. Injuries usually occur randomly and the times of occurrences are unpredictable except in a very large sample. Also, a single facility usually provides too small a statistical sample to be used as a control tool. However, occurrence of injuries has a direct correlation to situations of injury potential, particularly unsafe acts. Some studies show that, on the average, one loss of life or permanently disabling injury will occur for each 100 minor injuries or each 500 unsafe acts. Therefore, management must direct its effort toward eliminating high injury-potential situations. Eliminating the causes of injuries is, in the long run, the only way to reduce the occurrence of injuries. Illness statistics are not as discrete as injury statistics, so management must be even more alert to potential causes of occupational illnesses.

Essential Aspects of a Sound Safety and Health Program

The first step in creating a sound safety and health program is for the management team to personally accept the responsibility for safety and health. Facilities, including structural features, equipment, and the working environment, must be safe. Management must then insist that: (1) housekeeping be maintained at a high level; (2) safe working procedures be established; and (3) employees practice safe working procedures, being alert and applying good judgment to the hazardous aspects of all tasks, even to those not covered by formal safety and health procedures.

A facility that is safely constructed and maintained is a highly visible sign of management's commitment to safety and health. To have a safe physical plant, members of management must ensure that applicable physical safety standards, such as OSHA and MSHA, are met. Substandard facilities or equipment must be repaired promptly. Each employee's safety suggestion or complaint (the great majority of which will concern the safety of the physical plant) must be given its due consideration and prompt reply.

The Facility Manager must insist that employees achieve a high level of housekeeping and require that all work and travel areas be free of hazards, all materials and supplies be stored in an orderly way, and all materials and supplies be clearly identified.

Regular job training must incorporate safety and health principles. Operating and mechanical work procedures should include key safety points and sound practices. Management must establish safe work procedures for routine and hazardous jobs, such as entering vessels, working on energized equipment, and breaking into pipelines containing hazardous materials.

For a facility to attain its safety and health goal, employees must understand the facility's safety and health philosophy, follow good work practices, be alert to the hazardous aspects of the job, and use good safety and health judgment when undertaking new tasks. Communication, training, and involvement are essential to achieving a safety and health goal.

Responsibility and Accountability The Facility Manager is responsible for the safety and health of the facility. The manager is responsible not only to corporate management but also to the employees who have entrusted their safety and health to his or her direction. The execution of this responsibility is delegated to the staff, which must have the necessary authority to act.

The Facility Manager is accountable to the Corporate Manager for the safety and health performance of the facility. Safety and health performance is considered an important item in evaluating the person's capability as a manager. The Facility Manager, in turn, must hold all subordinates accountable for the safety and health performance of their departments.

In a large organization, the manager must delegate a major role in safety and health to the department heads. In smaller, less formal organizations, the manager will not need to delegate as extensively and should maintain closer personal control.

Since safety and health is a line organization function, the Facility Manager heads the Facility Management Safety Committee, the top of an interlocking committee structure extending through the complete line organization. The chairperson of the Management Safety Committee assures a continuing, vital, and stimulating safety and health program, encouraging the participation of the staff in planning and execution. Visible, sincere concern for safety and health must be shown to set an example for all members of management. Safety and health must be established as the first order of business, and subordinates must be required to accept this philosophy.

The manager's supporting role includes involvement with communications, feedback, evaluation, and planning. In the area of communications, the manager should be informed promptly of any serious occurrence. Safety and health must be the first order of business in any general staff meeting. The staff should make a brief safety and health status report to the manager the first thing in the morning or offer this report when an injury or unusual incident occurs, or when a safety and/or health accomplishment is made.

The manager must make regular, informal tours of the plant to observe safety and health conditions. When accompanied by the department head, the manager may make a more formal inspection of a specific department. Occasionally a formal or informal safety and health inspection of the plant should be made sometime other than on the day shift, Monday through Friday. Although not expected to participate in departmental meetings, the manager should attend them occasionally. Finally, the manager should make individual safety and health contacts with subordinates and hourly employees. During these personal contacts, the manager must avoid bypassing the line organization or usurping their responsibilities. Any correction or formal safety or health action must be done through the line organization.

Feedback is essentially upward communication from the line organization to the manager. It involves formal reporting, written and oral, as well as informal reporting of information resulting from personal contact with subordinates. In addition, statistical comparisons of injuries and unusual incidents can be used to develop information on trends. Unusual incidents are of particular value in this regard because they occur more frequently than injuries and, therefore, provide a broader sampling of the safety and health climate.

When evaluating supervisory safety and health performance, managers should consider the following indicators of good safety and health performance:

1. The supervisors maintain low injury and illness experience.

2. The supervisors maintain good housekeeping in each unit.
3. The supervisors approach safety and health problems positively and develop solutions and implement recommendations from safety and health professionals.
4. The supervisors cooperate with other groups in safety and health activities and projects.
5. The supervisors have original ideas for the safety and health program.

Indications of unsatisfactory safety and health performance include:

1. The supervisors quarrel with minor details of the safety and health program rather than carrying out the program.
2. The supervisors consider the safety and health program to be somebody else's program, rather than their own or the group's.
3. The supervisors tolerate unsafe acts from the employees.
4. The supervisors allow poor housekeeping in the area.
5. The supervisors fail to generate new ideas.

After evaluating the facility's safety and health performance, the manager can develop plans for safeguards and the elimination of exposures. The safety and health plans should be both immediate and long-range. Some examples of actions that can be incorporated into either short- or long-term plans as determined by the extent of the problem are:

- Correcting process hazards that developed through faulty design or frequent modification
- Developing a safety meeting topic schedule for use at departmental meetings
- Reviewing all safety work orders at staff meetings
- Providing an approved training course for all operators of powered industrial trucks or other mobile equipment

Intermediate Managers

Intermediate Managers Role in Safety and Health

While the overall policy and direction of the safety and health program must start with executive management, the effectiveness depends directly on the commitment and managerial skills of the line organization. Each member is actually a manager in his or her sphere of responsibility and is responsible for the safety and health of the personnel and the equipment assigned there.

Because safety and health is a good measure of performance, members of intermediate management should seize the opportunity of dealing effectively with safety and health to demonstrate their abilities as managers.

Goal of Intermediate Management's Safety and Health Program The general goals of an intermediate member of management are identical to those of the Facility Manager. In addition, specific objectives should be assigned to each member to support general goals. For example, the facility goal is to see that safe working procedures are established covering all frequently occurring or hazardous tasks. As an objective for his or her group, the department head, in conjunction with the safety and health department, might include a review and update of all written work procedures applying to the department.

Acceptable standards that include good housekeeping

and safe work procedures and practices within each department must be established to ensure a safe plant. It is important that these standards be developed with the participation of the line organization and safety and health department of the unit. Such participants will achieve a clear-cut understanding and acceptance of the safety and health program's goals. The use of a series of interlocking safety and health committees, starting with the Facility Manager's staff or the Management Safety Committee, provides for the participation of all levels.

After establishing acceptable levels for safety and health in the plant and in work practices, the intermediate manager must plan and implement activities to ensure compliance with these levels or standards on a continuing basis. Determining the degree of compliance requires a reasonable amount of time to ensure accuracy and thoroughness. Inspections, sampling-type observations, and other techniques may be used.

Deviations from the standards must be corrected and indicated corrections must be uniformly enforced and communicated, where applicable. For example, employees not wearing the required eye protection should be reminded of the unsafe practice, be told why compliance is important, and be informed that their compliance will be expected in the future. If three or four employees are found violating eye-protection rules throughout the department, the intermediate manager might discuss the eye-protection requirements in safety meetings with all members of the department.

Responsibility and Accountability The importance of personal example increases with the supervisor's visibility, reaching its peak at the level of first line supervision. Even more than the Facility Manager, the members of intermediate management must show their concern for safety and health, and this concern must be sincere and continuous. In addition to personal safe conduct on and off the job, safety and health must be set as the first order of business in all instances and subordinates must be required to do likewise.

Intermediate Management's Involvement in Safety and Health Each member of intermediate management must be personally involved in making decisions and executing programs. Participation brings pride, accomplishment, and commitment to reaching the objectives. Obviously, the intermediate manager must delegate responsibilities and action.

On the other hand, a reliance on "management by exception" can be used only to a limited degree. Delegation without feedback may be interpreted as lack of deep interest in the importance of safety and health. The member of intermediate management who demands a report on production results after each shift but who requests a report on housekeeping and unsafe practices only once a month weakens his or her safety and health effort. Subordinates will attach importance to safety and health goals and activities in proportion to the continuing interest shown them by their superiors.

For example, the Facility Safety Engineer reports on a safety inspection or evaluation. One way to handle the report is for the superintendent to send it down the line with a notation that it be read and the violations corrected. A much better way would be for the superintendent's staff to discuss both the general and specific items, agree on a program of corrective action, and set dates for implementation. Until the tasks are completed, each subordinate should report back periodically on the group's action and results.

The intermediate manager has more opportunity than the Facility Manager to make personal observations to check the validity of reports about the effectiveness of the unit's safety and health performance. However, the evaluation of performance like that of the Facility Manager must include upward communication from his or her organization on each objective of the program, personal observations, and independent audits by outsiders.

It is important for the intermediate manager to remember that management's safety and health activities serve primarily to strengthen the role of first line supervisors in their leadership of employees. The safety and health program must be evaluated constantly to see that it, too, does the same.

First Line Supervisors

The First Line Supervisor's Role in Safety and Health

All of the responsibilities and functions of the other members of management are related, in general, to first line supervisors. They are the contact through which the majority of management's communications flow. The following comments should be considered in a supervisory development program and in administering the safety and health program.

Goals of the First Line Supervisor's Safety and Health Program The first line supervisors' safety and health goals are specific: to see that safe working procedures are practiced, to maintain good housekeeping in their areas, to see that employees accept safety and health as their personal responsibility, and to insist on adherence to strict safety and health standards.

Responsibility and Accountability The first line supervisor must do many of the things that have been described previously for the Facility Manager and members of intermediate management. The following details may be of significant help to first line supervisors in meeting their responsibilities.

To encourage their groups' acceptance of safety and health goals, they must present the specific goals in group meetings and in individual contacts. This provides for two-way communication, with the supervisor acting as head of the discussion group. It is important that the group, led by the first line supervisor, be carefully incorporated into the interlocking safety committee program.

There is no substitute for periodic inspection to determine whether substandard physical conditions exist. Either the supervisor or assigned employees can make the inspections. Checklists can be helpful by including items that have special potential for injury, such as walking surfaces, ladders, guardrails, guards on moving parts of equipment, electrical equipment, and special safety equipment.

Maintaining a safe plant is absolutely necessary if the employee is to believe that management is sincere in its desire for good safety and health performance. Therefore, when substandard physical conditions are found, it is important that the areas be safeguarded until repairs are made. In addition, initiating orders or requests for repairs promptly and communicating the status of the item to employees are essential. The majority of employees' complaints or safety and health suggestions will relate to unsatisfactory physical conditions. It is important that members of supervision listen to each complaint or suggestion and give careful consideration to it, even if ill-founded. Giving a prompt answer and taking action where indicated are equally important.

The first line supervisor must specifically define the standard of good housekeeping in order for employees to understand and accept it. One possible standard is:

> An area must be clean and free from debris and unneeded materials. Materials and equipment needed

must be neatly stored in well-defined areas and in rows parallel to or perpendicular to fixed walls, roads, or other suitable reference lines. There must be no wire, hose, hole in pavement, or other obstacle or hazard in walking areas unless protected by warning signs or barricades.

Recognizing substandard housekeeping conditions requires frequent inspections by the supervisors and assigned employees. Involving employees in detecting and correcting unsatisfactory housekeeping conditions will result in better housekeeping, which will help prevent accidents. Maintaining satisfactory conditions daily is required if gains are to be made over a long period; periodic cleanup alone will not suffice. Housekeeping is a visible indication to management and hourly employees alike. The supervisor's standard of housekeeping has a major effect on his or her safety and health credibility.

Safe procedures must be established for all hazardous jobs. No one know the details of jobs better than the first line supervisor and the employees; therefore, their participation in either preparing or reviewing work procedures is essential. Since first line supervisors are the ones responsible for executing the work procedures safely, they are the ones in the best position to do the actual training and follow-up. However, they usually need instruction in how to train others.

The goal is for employees to understand the safety and health policies of the plant, practice good safety and health, be alert to the safety and health aspects of any job, and apply sound judgment to new tasks. Complying with established work procedures depends on three things: (1) the effectiveness of the employee's training; (2) the employee's interpretation of the expectations of his immediate supervisor; and (3) the supervisor's effective observation of his employee. Planned observations of work practices are essential to success. Supervisors must develop skill in observation and practice observing employees regularly in order to identify and correct unsafe work practices.

Deviations from safe work practices must always be corrected. Correction requires that the supervisor must first determine the reasons for the deviation and then act appropriately. Frequent causes for employees' deviation from safety and health rules include not accepting established procedures, not understanding methods, lacking the necessary skill, or valuing short cuts above the safety of the established method. Often, correction can take the form of reinstruction if the employee is convinced of the supervisor's sincerity and of the importance of compliance. The need for discipline is rare. In cases where it is necessary, it must be applied firmly and consistently.

Only through practice do employees become alert to the hazardous aspects of jobs. Alertness to hazards can be developed by discussing the key safety and health points for established jobs and by training and analyzing hazards for new assignments.

The following are examples of activities that increase an employee's level of safety and health awareness:

1. Help to set standard work procedures and periodically review them to keep them up to date.

2. Discuss safety and health matters with other employees in safety committee meetings led by the supervisor.

3. Carry out inspections and report on safety and health equipment, housekeeping, unsafe conditions in the plant, hazards uncovered by an injury in another department, and so on.

4. Submit either oral or written safety and health suggestions. Publicizing the acceptance of successful suggestions provides recognition to the employee.

5. Have management recognize outstanding group performance.

6. Have the supervisor listen to employees sincerely and with respect for their points of view.

The First Line Supervisor's Safety and Health Involvement
To the employee, first line supervisors are the most important members of management because they are the members of management directly in contact with the employee and because they constantly make decisions that directly affect what the employee does. Therefore, first line supervisors' personal examples—their safe practices, their interest in the safety and health goals, and their insistence or lack of insistence on high safety and health standards—determine almost entirely the attitudes and performance of their employees. Their influence makes or breaks the safety and health program.

Unless first line supervisors feel that they are full-fledged members of the management team, their performance in safety and health may be seriously hampered. In most organizations, the authority of foremen is necessarily limited on major matters to initiation of requests or recommendations for action. This is particularly true of major expenditures, major disciplinary actions, the selection and assignment of employees, and the like. They may feel that they lack the authority commensurate with their responsibilities unless they participate in planning and decision making. Management should seek their recommendations on matters that affect their responsibilities. In addition, first line supervisors may be particularly sensitive if bypassed by higher members of the line organization who deal directly with their employees. Every effort should be made to work through first line supervisors and, where that is impossible, to keep them well informed.

First line supervisors are generally promoted employees with limited formal education compared to others in the management structure. It is important that they receive training in how to carry out their duties, including such things as conducting safety meetings, doing the necessary recordkeeping, making housekeeping inspections, conducting job safety analyses, and giving job training to their employees.

Because of their interfacing position between management and the employees, first line supervisors may be burdened with a heavy flow of communications, paper work, follow-up, repair, training, and miscellaneous duties. Unless their work load is carefully controlled, they may feel overburdened and frustrated. This is particularly true regarding the volume of paper work. If supervisors are overloaded, they will devote their efforts only to those areas of responsibility that they believe their superiors consider urgent. Supervisors will forgo safety and health for production if they believe getting out the product is really all the management wants.

These examples are written for larger corporate structures with several layers of management. Your company may not be structured in the same manner, but rearranging the responsibilities to fit your particular structure will produce the same desired effect.

Summary

In order for your safety and health program to succeed, you need to delegate responsibility to specific jobs or departments and to various levels of your company. This can be done in the following ways:

1. Review your existing structure.
2. Determine what part each position should have in

carrying out the safety and health program and the levels of authority and resources necessary to do the job.

3. Determine the responsibilities that must be carried out by each position, so that each person can contribute to an integrated safety and health program.

4. Communicate with the employees involved, and discuss the responsibilities and authority in face-to-face meetings.

6

Developing Accountability for Safety and Health

Why develop an accountability program? Is accountability not inherent in hiring and firing, raises and bonuses? What is the purpose of such a program, and why is it important? A good way to understand its importance and purpose is through the following example.

Imagine a football club with an owner, manager, coach, and team of players. You can easily see that each person has a specific and critical part to play in the overall success of the team. The owner must see that operating capital and other necessary resources are available, and must oversee business matters in general. All final decisions must be cleared through the owner. Often the owner has the largest stake in profits or losses incurred by the team. The manager, on the other hand, is responsible for implementing major decisions and policies set forth by the owner, for developing and implementing overall strategy, and for reporting back to the owner on new and changing resource needs. The coach has many roles, including teacher, friend, supervisor, coordinator, and disciplinarian. He must take policy decisions made by upper management and see that these policies are carried out daily. He is responsible for each game plan. He must also interact closely with individuals and specific groups of individuals on the team as needed. While the team is playing, he stands on the sidelines making sure that the team is doing what he has taught it to do, following the plan he has devised. The players do the jobs of kicking, passing, and carrying the ball to make touchdowns. Within the team, each member has his own area of responsibility, such as quarterbacking, blocking, or running. But the whole team suffers when any player performs poorly.

An accountability program is a system that ensures that all people on the team fulfill their responsibilities. When players fail to show up for practice with no reasonable cause, they are fined. Players who perform poorly, for whatever reason, are not in the starting lineup. Player contracts reflect trends in poor performance or relative value to the team, thus creating a form of personal accountability for performance. We have all heard of coaches or managers fired at the end of a season, and sometimes even in the middle of a poor season. Such actions create a very real sense of personal accountability among coaches and managers. For owners, profit and loss considerations are a powerful motivation. Such concepts as reputation are also strong motivators for all team members. Thus you can readily see the purposes an accountability system serves for a ball club, and how important it is.

Business also involves owners, managers, coaches (or supervisors), and players (or employees), and each person on the team has his or her area of responsibility. Unfortunately, in a business—particularly a small business—these areas are not always clearly defined. The team members may not understand that each person must perform at top efficiency in order to create a successful team.

Sometimes the owner may also function as manager and/or supervisor. This dual responsibility may create problems when new initiatives are started if no additional personnel are available to support existing programs. Supervisors are sometimes asked to double as managers or workers as needs arise. These kinds of flexible and undefined (yet often necessary) organizational structures in a small business can lead to breakdowns in accountability systems.

In large businesses, responsibilities are frequently so complex that some get overlooked, causing a breakdown in accountability systems. Accountability also breaks down when responsibilities are assigned without needed authority or resources.

The purpose of an accountability program is to help *all* team members understand how critical their performance is, and to teach them to take personal responsibility for their performance. The importance of this task for a company safety and health program is to ensure that your safety and health program is not just a "paper tiger" with no real power to win (i.e., to achieve its objectives). Here are the steps to assuring accountability.

Set a Clear Goal and Assign Responsibilities

Before you can hold people accountable for their actions, you must be sure they know what their jobs include. They must have goals set for their personal performance. The objectives must be understandable, measurable, and achievable. It must be clearly established who is responsible for performing what tasks. Check your assignments of responsibility again to make sure that they clearly state who does what, and that all the objectives you plan to accomplish can be achieved. A hazy concept of responsibility will allow everyone to pass the proverbial buck, and you may be unable to determine who failed to perform satisfactorily. Assigning responsibilities to individuals without delegating authority to them or committing sufficient resources will set the scene for failure. A diligent worker will become demoralized; others will find ways to avoid the responsibility. Either situation will result in objectives not being accomplished.

Set Individual Objectives for Accountability Systems

Objectives for individuals should be based upon performance measures, that is, indicators that tell you whether the person either did or did not perform as expected. When setting objectives, remember the following:

1. Objectives should be aimed at specific areas of performance that can be measured or verified. Example: An objective "to improve safety and health performance in my department next month" is too general to be useful. A better objective is "to decrease first aid injuries by 10 percent next month." Even more measurable objectives are those for which the manager or supervisor has complete control, such as "to hold half-hour safety meetings for all employees in my work-group every Monday morning."

2. Objectives should be realistic and attainable but represent a significant challenge.

- Appropriate authority is necessary. Example: A personnel manager or safety director's objective to improve the safety and health record in the Press Department is not directly attainable since successful achievement would be dependent on the performance of the Press Department supervisor and the personnel that he or she supervises. A personnel manager or safety director's objective to determine specific classroom safety and health training needs, locate or develop the training, and notify managers of its availability would be within the bounds of his or her authority, and therefore achievable.
- Adequate training is necessary. Example: A supervisor's objective is to investigate all accidents and near-misses that occur in his or her area and ensure future prevention. This objective is unattainable if the supervisor has not been trained in accident/incident investigation techniques and appropriate hazard recognition. The supervisor will also need training in or need to have access to knowledge of appropriate hazard correction technology.
- Appropriate resources are necessary. Example: If a maintenance manager's objective is to ensure that all machinery is safe to operate, that objective will be unattainable unless the appropriate budget for replacement parts and capital improvements is adopted. If the manager is held accountable for a clean area at the end of each shift but is not given enough personnel to finish the cleanup before the next shift starts, he or she will find that an objective of clear aisles and work areas by the start of each shift will be unattainable.

3. Objectives should be mutually understood by both sides. Use language that has clear meaning to all those directly concerned and leaves no doubt about what each person is supposed to do.

- Example: An objective "to investigate accidents to determine multiple causation" may mean little to a supervisor. "Investigate accidents to determine all causes and take corrective action within 24 hours of the accident" is a clearer, more specific objective.

4. Objectives should be mutually agreed to by both sides. Even when you and your supervisors agree on most issues, you should discuss your managers' and supervisors' safety and health performance objectives with them and secure their agreement or cooperation.

Writing Objectives

Put each objective in writing. This will help you clarify your meaning, signal that you are serious about meeting the objective, and enable you and others involved to refer to the objective when a question arises. In specific terms, state what is to be achieved, to what degree, and the deadline for achieving it. Focus the terms you choose on performance. Give a copy of the performance objectives to the employee for whom they were written and use them in performance discussions with that employee. Following are some examples.

1. Conduct weekly inspections of my department with emphasis on housekeeping, personal protective equipment, preventive maintenance, and the wear of critical machine parts.

2. Determine the cause(s) of any accident occurring in my department and take corrective action within 24 hours.

3. Track to elimination all hazards identified through employee reports of hazards, accident/incident investigations, and/or weekly planned inspections.
4. Complete one job safety analysis each month for my department.

Reviewing Objectives

Review the performance objectives periodically to make sure you are getting the desired performance and results. For instance, if a supervisor met the objectives but the department continues to have too many accidents, too many close calls, too many unsafe acts, or no improvement in conditions, then the objectives need to be reviewed and revised.

Performance evaluation can be oral, written, or both. The critical elements of an effective evaluation include the following:

1. The evaluation should be performed at specified intervals. If performance evaluation is new to your business, the intervals should be short in the beginning so that unacceptable performance can be changed quickly with minimal effort. As the individuals become accustomed to their newly defined tasks, the intervals can be lengthened, using the evaluations for encouragement and refresher training sessions.

2. The evaluation should always be performed against a backdrop of previously defined objectives (as discussed above), so the person being evaluated will encounter no surprises concerning what was expected of him or her. Should problems develop, it may be necessary to modify the objectives to ensure that they are achievable and measurable. If it becomes clear that the objectives were not understood, "retraining" will be necessary.

3. The attitude of the evaluator and the person evaluated should be one of exploring ways to improve both the system and the performance of the individual. Attitudes such as refusal to listen to one another, animosity, blaming others, fear, or intimidation will not be conducive to achieving the desired results.

4. The goal of the evaluation session should be to encourage personal responsibility and the individual's ability to improve the performance of the "team." During the evaluation, give positive reinforcement for a job well done. This can range from verbal expressions of gratitude or commendation to more tangible rewards such as bonuses, awards, raises, and so on.

5. Both parties must be able to come to some agreement on needed changes in objectives or performance. If the evaluation identifies that performance did not meet its objectives, some changes must be made. In some instances the required changes will be obvious; in other cases the changes may need to be carefully explored. Perhaps the wrong person was assigned a particular responsibility, and a simple change in assignments may alleviate the problem. Or the level of authority of the assigned person may need to be increased in order for him or her to carry out the responsibility efficiently. In other cases the objectives themselves may need to be modified to allow individuals to develop capabilities they do not at present possess, and for which, therefore, they should not be held accountable.

6. The agreed-on changes must be incorporated into any set of performance objectives already developed. Many evaluation systems break down when managers fail to incorporate and implement identified items for change into existing objectives.

7. There must be a point where some predetermined

consequence for poor performance begins. Keep in mind that some task monitoring may be necessary to support the performance evaluation. For example, monitoring of a supervisor's ability to perform accident investigations may be needed after each accident until it is clear that the supervisor has learned the necessary skills. This task monitoring may then form the substance of performance evaluations done at a later date. Keep in mind that the complexity and formality of your evaluations should match the complexity and formality of the rest of your safety and health program.

SET CONSEQUENCES FOR FAILURE TO PERFORM ADEQUATELY

Initially, consequences for poor performance should be nil or minimal as the person learns new skills and changes behavior patterns. Instead, use positive reinforcement during this initial phase of performance evaluation to encourage the natural desire of most employees to feel successful.

Although the goal of any accountability program should be to develop a sense of personal accountability for actions, often individuals need to know there are negative consequences for poor performance. The purpose of the consequences is to reinforce the importance of meeting objectives. It is also critical that supervisors and managers understand *when* the consequence will occur so that there is no element of surprise and subsequent confusion on their part.

The severity of the consequence should match the degree of underachievement. Firing a supervisor for the first poorly conducted accident investigation is an obvious example of a consequence that does not fit the problem.

Gradually, though, the consequences for poor performance should be increased to some specified maximum level. One common example of such gradation is this disciplinary system: (1) oral warnings; (2) written warnings; (3) fines or suspensions; and finally (4) termination. However, you may find that other consequences may produce the desired result. You can experiment with a variety of consequences as long as your employees are fully informed of your intentions.

If poor performance continues, at some specified point you may conclude that the individual is not capable of handling the assigned responsibilities. If sufficient training and nurturing through the accountability system has been documented, the reason for the problem (inadequate capabilities, improper attitudes, or the like) should not be the issue. The maximum level of consequence must be enforced. Otherwise other employees will believe that consequences are not important or are enforced inequitably, and such a belief will block any chance for an effective accountability program.

Summary

An accountability system is essential if all the hard work and effort you spent in developing a safety and health program is not to be lost. However, the concept of an accountability program is more than enforcing punishment for "bad" employees (including managers and supervisors). The goal of the accountability program is smoothly and efficiently to teach your managers and supervisors to take personal responsibility for their actions and subsequent effect of these actions on the "team." This is achieved by clearly defining their expected performance, allowing them the freedom to learn in a nonthreatening atmosphere, and enforcing negative consequences only when training and nurturing have not been effective. The employees must have a clear understanding of the type of, severity of, and time when the consequences will be enforced. This interaction between the employer and employees will allow the employees the opportunity to choose for themselves the outcome: whether to change performance, admit to an inability to perform adequately, or endure the consequences.

7 Safety and Health Quality Control: Determining How Well Your Safety and Health Program Works

You know your safety and health program is in place. You have set your safety and health goal for the year and clearly stated the objectives, procedures, and activities necessary to meet that goal. Responsibilities have been defined and clearly assigned, and adequate authority and resources have been allocated to carry out the responsibilities. People have been trained in their roles in the safety and health program, and they clearly know the consequences if they do not carry out their assigned responsibilities.

But your responsibility for employee safety and health is not fully discharged. The next critical step is to evaluate how well your safety and health program is working to achieve your goal. This process is not an inspection. Inspections are necessary to look at the facility, the process, and each job to see if any hazards exist so that they may be eliminated or controlled. This process is also more than the usual audit. For example, if you are evaluating employee participation by looking at the activities of the safety committee, you will want to know if that committee met at the intervals specified, and if most of the members attended each meeting. These are audit questions. But you want to move beyond this simple accounting to ask if and how employee participation at these meetings helped improve the safety and health program at your site. How is the work of the safety committee helping you meet your goal? These are evaluation issues. Interviewing the committee members might lead you to some answers. Asking supervisors whether activities of the committee have changed employee behavior or changed attitudes in the workplace such that safety and heath are improved will also help the evaluator judge the committee's effectiveness.

A safety and health evaluation looks at the systems you have created to carry out your safety and health program to see if the systems are working efficiently to enable you to reach your goal.

Three tools are useful for this evaluation: (1) document review, (2) employee interviews, and (3) review of site conditions. Although evaluations can be done by people at the site, the best evaluations are done by people who are knowledgeable about the process(es) at the site and about managing safety and health programs but who are not working at the site being evaluated. The fresh look an outsider brings produces a more accurate and helpful evaluation. This outsider may come from corporate headquarters or another worksite within the company, an insurance company, or a consulting firm.

All systems that contribute to your safety and health program should be reviewed. These should include management leadership and the evaluation of that leadership, the analysis of your worksite to identify hazards, hazard prevention and control, accident and near-miss investigations, employee involvement, safety and health training, use of personal protection equipment, and adequacy of the health program and the emergency response program. Your site may have additional programs or systems that contribute to the safety and health program that you need to evaluate.

Evaluation often causes anxiety among the people at the site to be evaluated. When planning the evaluation, if you clarify that the evaluator's focus will be on the systems rather than on the personnel, some of that anxiety may be dissipated.

The final report should be a narrative account, not a check list. Some people include in the report pictures of both situations that are excellent and of those that must be improved. Some provide a grading system so that each year's results can be quickly compared to previous years. But the body of the report should list the programs or systems reviewed and a narrative report of the examination of each system or program. The report should contain a schedule for when changes will be completed, who is responsible for seeing that the changes are made, and when the changes were actually completed. This report should be available to any employee who wants to read it.

For safety and health to remain a visible top priority at any site, good management programs must be in place, but these programs should be examined yearly to be sure that they continue to contribute to meeting the goal set for that year, and to make the changes necessary to point the programs toward the goal set for the next year.

The material in this chapter will explain in detail, with examples, just how this examination can be done effectively.

What Should Be Evaluated?

Ideally everything that you know to be directly contributing to your safety and health program should be evaluated. Since OSHA has determined and published the elements of a good safety and health program as the OSHA Safety and Health Program Management Guidelines (54 FR 3908, January 26, 1989), the outline of that publication should help you determine areas to be covered by your evaluation. The outline is as follows:

I. The demonstration of management leadership through:

A. Setting and communicating the safety and health policy

B. Setting and communicating a clear goal and objectives

C. Being visibly involved in employee safety and health

D. Assuring employee involvement in safety and health problem identification and resolution

E. Clearly assigning responsibility for safety and health

F. Giving adequate authority and assuring efficient use of resources

G. Holding line supervisors accountable

H. Assuring quality

II. The analysis of your worksite to identify hazards through:

A. Comprehensive safety and health hazard surveys

B. Analysis of planned changes to identify hazards they might introduce

C. Routine hazard analyses such as:

1. Job hazard analysis (also known as job safety analysis)
2. Process hazards analysis (used in industries with complex and hazardous processes)
3. Activity hazard analysis (mostly used in construction

D. Employee reports of hazards

E. Accident/incident investigations

F. Analysis of injury/illness trends

III. Hazard prevention and control techniques through:

A. Engineering controls

B. Work practice controls

C. Personal protective equipment

D. Administrative controls, when necessary

E. Disciplinary systems to enforce controls

F. Preventive maintenance

G. Emergency preparedness

H. Medical program

IV. Providing safety and health training to ensure that your employees at all levels know how to protect themselves and others from the hazards and potential hazards of your worksite

Who Should Evaluate?

The person or persons selected to do a safety and health program evaluation can be drawn from the workplace safety and health department or the safety committee. But the best evaluators will be people not involved with the day-to-day operations of the site. Sources for finding an evaluator with fresh eyes for the site are the corporate safety department, a knowledgeable person from another site in the company, insurance companies, and consultants. It is sometimes helpful to have two activity managers switch and evaluate each other's results.

Tools for Collecting Information Used in Evaluation

There are three indispensable evaluation tools for judging the effectiveness of occupational safety and health program management:

- Review of written programs and records of activity
- Interviews with employees at different levels
- Review of site conditions

Documentation

Every worksite will have, at an absolute minimum, written accident reports and the OSHA log of injury and illnesses as required by law. Major companies should have written procedures and records of all their safety and health programs. The evaluator should compare the written program to the written records of what occurred.

Interviews

In addition to the documentation or where it is not (fully) available, interviews can be very helpful as a second source to establish what has occurred. There are two basic kinds of interviews: formal and informal. The formal interviews are conducted with randomly selected employees who are asked preselected questions. Informal interviews occur at employee work stations and loosely follow a list of topics.

To assess how well the worksite safety and health policy is communicated and understood and how well the disciplinary system works, ask the employees to explain them.

To understand how well safety and health training has succeeded, interview hourly employees and first line supervisors. Ask employees to describe what hazards they are exposed to and how they are protected. Ask them to explain what they are supposed to do in several different types of emergencies. Ask supervisors how they teach, how they reinforce the teaching, how they enforce safety and health rules and safe work practices, and what their responsibilities are during emergency situations.

Interviews with management about safety and health program activities should focus on management's involvement and commitment to the safety and health program. Ask how the policy statement was created, and how that statement is communicated to all employees. Ask what information management receives about the safety and health activities and what action management takes as a result of that information. Ask how management commitment to safety and health is visible to the workforce.

Review of Site Conditions

The conditions at the worksite reveal much about the safety and health program's effectiveness. Worksite conditions can be seen indirectly by looking at documents such as inspection reports of hazards, employee reports of hazards, and accident/incident investigations.

Site tours may also reveal hazards. Be careful, however, that the site tour does not become an inspection, with emphasis on hazard correction. When a hazard is found, in addition to assuring that it is corrected, ask what management system should have prevented or controlled the hazard. Determine why that system failed, and also correct the system.

Evaluating Safety and Health Program Activities

Time and resources can be squandered when safety and health program activities do not achieve the desired results. Each year, therefore, activities should be planned to achieve objectives that lead to the goal set for that year's safety and health program. The yearly evaluation assesses that achievement.

Plan = Goal → Objectives → Procedures → Activities

To begin an evaluation you reverse the process, comparing program activities and the procedures that govern them to program objectives. Then evaluate the objectives against the goal of the program. If this analysis finds program efforts that are not effective, recommendations for changes in the program can be made for the next year.

Evaluation = Activities → Procedures → Objectives → Goal

Activities and Procedures

Do the actual safety and health program activities and the procedures for organizing them bring the results expected when the safety and health program was written?

Evaluation = **Activities** → **Procedures** → Objectives → Goal

Smaller Worksites

Even if yours is a smaller business, you should stop and think about the manner in which safety and health activities were carried out for the period evaluated and whether the results achieved were those expected at the outset.

Larger Worksites

Large companies will have written procedures for the major activities of their safety and health program. They will also have written records of those activities as they were performed. Evaluating whether or not the written procedures were followed in the period evaluated or how well they were followed is an audit function of quality assurance or program evaluation.

Sample Questions

The precise questions you should ask and get answered during the audit portion of the evaluation will depend upon the activity audited and the way it was planned to be accomplished. For example, if a certain person or group of people are to carry out the inspection program, were they the ones who actually did the inspections? (In many workplaces, inspections are conducted by the person with the most expertise and members of the site safety and health committee.) Was the "expert" on every inspection? Did the employee members always participate?

Other questions about inspections might include the following: Is there evidence that the "inspectors" went to every part of the worksite that was planned for each inspection? Did their reports indicate that they were finding the kinds of hazards that they were trained to recognize? Did hazard correction get appropriately assigned? Did the hazards get corrected in an appropriate and timely manner? Was the correction tracked to completion?

Similar questions should be asked and answered about each activity under the safety and health program. When a difference is found between the way the activity was planned and the way it is being carried out, assess which way best meets the objectives and goal. Then make sure that everyone follows that procedure.

Objectives

The objectives connect the goal for the safety and health program to the program procedures and activities.

Evaluation = Activities → Procedures → **Objectives** → Goal

Objectives That Can Be "Audited"

Sometimes, when setting up new activities or improving existing ones, one or more of the objectives will be to complete these activities. For example, suppose the objective is "to complete one job safety analysis *each month,* with follow-up revision of safe work procedures and employee training in the *following month.*" In this case, the objective describes the *frequency* of activities rather than the result.

Because the question is simply whether the activities occurred, not what is the *quality* of the results, the evaluation would be more of an audit. Since the objective is being audited, look for evidence that job safety analyses were done each month. Is there evidence that revisions of procedures and training were also made each month following the job safety analysis that was done the month before? If the answer is yes, and other program evaluation reveals no need to do something different, the frequency of these activities would become one of the ongoing procedures to be audited.

Objectives That Must Be "Evaluated"

Ordinarily, objectives should be aimed at the results desired from the program activities. For example, you might have an objective "to identify and assign all areas of safety and health responsibility that are not now clearly assigned, so that all safety and health responsibilities can be carried out successfully." There would then be a set of activities needed to accomplish that objective. The activities might include assigning a committee to implement the objective by listing all the safety and health responsibilities; reviewing assignment of those responsibilities; identifying missing assignments, duplicate assignments, and unclear assignments; and recommending clear assignments.

In this case the evaluation would focus on whether the objective was accomplished. In other words, were all areas of safety and health responsibilities that were not clearly defined actually identified and clarified? Perhaps these areas were identified and corrected, but the corrections have not been made because the Personnel Department still must rewrite seven major job descriptions to get the responsibilities correctly assigned. This then leads to a new objective, to get the seven job descriptions rewritten with clearer assignment of safety and health responsibilities. If all areas were identified and clarified, this objective would not lead to another objective. To take another example, if the objective is to increase employee safety and health awareness by involving the employees' families in a safety and health awareness program, then the evaluation would begin with some sort of measurement of the difference between employee awareness before the family program and after it. One way to measure that difference might be to ask the employees who participated in the program to complete a questionnaire; another would be to select employees randomly to ask if they thought their awareness was raised. Still another would be to ask supervisors if they saw a difference in employee awareness after the family program began. If, after using some type of meaningful measurement tool, the findings were that the program had increased awareness, the family program would be an ongoing activity subject to audit. If the findings were that no increase could be measured, a different family program could be tried to implement the same objective, or a different objective could be set to increase employee awareness.

Goal

Evaluation = Activities → Procedures → Objectives → **Goal**

The goal provides the framework for the safety and health program. Objectives should be evaluated to be sure they are leading to the program goal. For example, suppose the goal was the reduction of employee exposure to hazards, and one of the objectives to achieve that goal was to hold safety meetings for all employees every month. Since this is an activity objective, it could be audited to make sure that safety meetings were held. But the next question is "Did meeting this objective help meet the goal of the safety and health program?" In other words, did the meetings help employees understand the hazards to which they are exposed and result in plans to reduce that exposure? Here are several ways to collect that information. If employees are asked to sign off on a sheet to show that they attended the training, they could also be asked to describe the hazards and potential hazards that exist in their work area, and how to control them better. If there was no sign-off sheet, interviews with some randomly selected employees could reveal their assessment of whether the meetings helped their understanding of the hazards and resulted in plans to control them. Interviews with supervisors would also show whether employees showed better understanding after training.

If the results indicate that not much was learned at these

sessions, ask further questions to see what went wrong. In this manner, all the objectives should be checked to see if they helped achieve the goal. Stopping to evaluate programs against the goal of the safety and health program can avoid wasting time on an activity that is not really helping to make the worksite a safer, more healthful place to work.

Evaluation Judgments

Gathering information about activities under the safety and health program is very important and is the most time-consuming part of program evaluation. It is, however, the easiest to understand and to do. The hardest part is making judgments about program effectiveness. Without good judgments, the evaluation becomes a waste of time. With good judgments, it can be the most important part of the safety and health program.

Some sample sets of questions shown in the sections above can be used to guide the evaluator to judgments. There is also a sample section of evaluation instructions shown in Appendix 3. The evaluation should be viewed as determining the "bottom line" in the sense of the program goal "profit." The "profit" in program activity is any contribution to the achievement of your goal. All of the decisions that evaluators arrive at should be driven by the concept "What does it do to contribute to the goal?" To assist businesses in making these judgments, the sample check list in Appendix 3 can be used to help decide what information should be gathered and what judgments need to be made.

Worksites should draw up their own specific procedures for gathering information and making judgments. They may add environmental, product safety, or security aspects to the system as well. The questions that need to be asked to guide the judgment should be based on individual site program activities and site objectives.

How to Use the Evaluation

The evaluation will only prove valuable if it results in improvements in meeting the safety and health goal. Some of the recommendations that result from the evaluation will be for one-time corrections. Many, however, will involve changing emphasis or trying new program activities. These recommendations should be incorporated into the objectives for the next year. Think about setting up one "permanent" objective "to audit the procedures set for safety and health program activities."

Larger Worksites

The evaluation should result in a written report with written recommendations and documented follow-up to those recommendations. It may be useful to refer to past years' evaluations when preparing new ones or when planning new objectives. If you find that the same recommendations are being made year after year, the system of tracking recommendations to completion needs improvement.

Smaller Worksites

At smaller sites where a written evaluation report may not be practical, make sure you set aside time to think about what you want to change. You have thought about what you did during the course of the year; you have talked to people; you have looked at your working conditions; and you have reviewed the documents that you do keep. Now decide what you want to do differently, and make sure that everyone understands what is expected from now on.

Starting or Overhauling a Self-Evaluation System

Establish the Areas to Be Evaluated

To decide what needs to be evaluated, start by looking at the list of items in the section above called "What Should Be Evaluated." The sample check list in Appendix 3 can also serve as the core of the things you should want to evaluate. Small businesses may not need to add much to that group.

Larger businesses need more detail. For example, the check list for shows only one indicator for routine inspections; i.e., "routine inspections are conducted regularly by adequately trained personnel." Larger worksites will want to spell out more detail, such as whether inspections follow the procedures set for them, whether they are finding the hazards they should find, whether the hazards are tracked to correction, or whether the same hazards keep reoccurring. These thoughts should also be part of the evaluation in a small business, but they would not necessarily need to be spelled out so clearly. Be sure that all objectives set for your safety and health program are included in the evaluation.

Select the Evaluator

The evaluator should be familiar with the activities used in the safety and health program but not be intimately involved in overseeing or implementing them. He or she should know safety and health program management.

Use Self-Evaluation Tools

Make sure that whoever does the evaluation knows how to gather the information to be used in making the evaluation judgments. Help is provided in Appendix 3.

Provide Instructions to Evaluate Activities Against Procedures, Objectives, and Goal

Make sure that whoever does the evaluation knows that the activities should be evaluated against procedures, objectives, and the goal of the program. A good evaluation report format can help you accomplish that. (See the example of a section of a format for a narrative report in Appendix 3.)

Insist that the evaluator find the "bottom line"; that is, what has this activity done to contribute to the goal? An evaluation that provides no insight into the program should be rejected. In particular, do not accept a narrative that only describes the programs as they are designed. Insist that the hard questions about the effectiveness of the program be asked and answered.

Use the Evaluation to Determine New Objectives

When the evaluation is complete, you should have a list of new items that you wish to make objectives. Check last year's objectives to see which of them need to be continued and which are either completed or do not help reach your goal.

Summary

This chapter has explained what a safety and health audit and evaluation is and how to accomplish this task. The chapter describes what should be evaluated, who should do the evaluation and with what tools, how the evaluation should be done, how to use results of the evaluation, and how to begin or to overhaul the evaluation system you are using.

Appendix 3 offers an example of a self-evaluation check list that may be used to perform a comprehensive evaluation of the safety and health program. The check list serves as a step-by-step guide to using the three evaluation tools—documentation review, employee interviews, and hazardous conditions review and analysis—to assess each element and subelement of your safety and health program. It also provides a sample evaluation instruction sheet that can be used to evaluate routine inspections when such inspections are part of your safety and health program.

Using this information to evaluate your safety and health program annually, you will be able to see clearly the bottom line in safety and health matters just as you now can scrutinize the bottom line in financial matters. In smaller companies, plant management will have the information they need to provide quality safety and health protection to all employees. In larger companies, by reporting this information to Boards of Directors, management at the highest levels can make knowledgeable and effective decisions to improve workers' safety and health. Corporate leaders, closer to the employees than OSHA, can lead the way to ensuring safe and healthful working conditions for America's working men and women.

8

Establishing Complete Hazard Inventories

If you want to protect your employees from hazards, you must understand just what those hazards are. The law requires that you provide a workplace free from recognized hazards that could cause death or serious injury. Are you sure you know all of the generally recognized hazards that could be part of your workplace or your working conditions?

What you need is a systematic way to compile an inventory of potential hazards in your workplace. The Safety and Health Management Guidelines were developed to provide guidance on how to manage workplace employee protection. These guidelines list three separate actions to take to identify all workplace hazards:

- Periodic comprehensive safety, industrial hygiene, and health surveys

- "Change analysis" of the potential hazards in new facilities, equipment, materials, and processes
- Routine hazard analysis such as job safety analysis, process hazards analysis, or activity hazard analysis

These three major actions—comprehensive surveys, change analysis, and routine hazard analysis—form the basis from which good hazard prevention and control can be planned. After controls are put in place, additional worksite analysis tools such as inspections, employee reports of hazards, accident and incident investigations, and accident pattern analysis are used to ensure that the controls stay in place and that other hazards do not appear.

But first, you need to understand the hazards and potential hazards in your worksite.

Comprehensive Surveys

Comprehensive Surveys Compared to Inspections

An inspection is often done by employees at the site. Comprehensive surveys should be performed by persons whose knowledge is both broad (covering more than just your industry) and deep. You will want to find people who can bring fresh eyes and extensive and intensive knowledge of safety, health, or industrial hygiene to your worksite. Because there are few professional consultants who are equipped to do comprehensive surveys for industrial hygiene, occupational health, and safety, the best approach is to use a team consisting of three separate specialists: a safety professional, an industrial hygienist, and an occupational health professional. The occupational health professional may be a physician or a registered nurse with specialized training and experience in occupational health. This person may assist the safety professional or the industrial hygiene professional, or may do a separate health survey, depending on the circumstances at your site.

For small businesses, safety and industrial hygiene experts can usually be found in the OSHA-funded, state-run consultation service. Occupational health professionals can sometimes be found at local clinics or hospitals. Larger businesses may contract for safety and health expertise or find it at the company or corporate level. Occupational health expertise may be no farther away than the plant nurse.

If you use any experts from within your company, be on guard for "tunnel vision," which can result in a failure to recognize hazards in areas outside the mainstream of your industry. You should strive to ensure that your maintenance shop is just as safe as your production lines. OSHA frequently finds unguarded saws and grinders, noncode electrical wiring, and similar basic safety hazards in areas outside of the main production process but which are being used regularly by employees.

For the industrial hygiene survey, at a minimum, inventory all chemical and hazardous materials in the plant and review the hazard communication program. Many industries need noise-level surveys and a review of the respirator program.

Questions to Ask Before Contracting for a Survey

To ensure that your worksite will receive the sort of comprehensive survey envisioned by the guidelines, you can ask potential surveyors certain questions before you engage them to do the survey.

1. What type of training has your prospective surveyor had?

A. How recent is it?

B. Is it too concentrated in your industry or based only on *experience* in your industry?

C. If he or she has a professional certification, is it still valid or has it lapsed for lack of recent training or seminar attendance?

2. What kind of information will he or she need in advance? A professional who is planning an in-depth survey will prepare for it by finding out as much as possible about your worksite and its processes before arrival.

A. Both safety and industrial hygiene professionals will probably want to see a layout of your operations.

B. The industrial hygienist may ask for a list of the chemicals you use, or the Material Safety Data Sheets (MSDS) you have received from your suppliers, and the types of processes in which you use them.

3. What kind of test equipment will the consultant bring?

A. You should expect the safety professional to bring:

- A tape measure
- A tic tracer (or similar equipment) to check wires or electrical equipment to see if they are energized
- A ground fault circuit interrupter tester

B. The industrial hygienist should bring noise testing equipment and sampling pumps or grab sampling devices, depending upon the chemicals or other contaminants expected.

4. How long will the survey take?
   - A. It should take several times longer than a routine inspection of your worksite.
   - B. If the industrial hygienist does sampling, it should be time-weighted, eight-hour or full-shift sampling to understand the overall exposure to employees.
5. Ask for references and check those where comprehensive surveys have been done recently.
   - A. Ask if any OSHA inspections occurred after the survey and, if so, if there were any serious hazards that the consultant missed.
   - B. Find out what kind of tools the consultant brought with him or her and what the survey covered. (See item 3 above and items 1 through 3 below.)

Gauging the Thoroughness of the Surveyor's Job

Even if you have someone do a "comprehensive survey," it will not provide all you need to know if it is not done thoroughly. Here are some of the signs of a thorough survey:

1. Safety professionals, industrial hygienists, and occupational health professionals should start with your injury and illness logs, looking for patterns.
   - A. The safety professional may also want to see other program documentation.
   - B. The Industrial Hygienist and occupational health professional will want to see your hazard com-

munication program and, if applicable, your hearing conservation and/or respirator programs.

C. The occupational health professional will want to see your records of employee visits to clinics and first aid stations and other sites where they receive treatment for work-related illness and injury. He or she will want to see records of employee training for first aid, CPR, and EMT. Baseline and follow-up testing records will probably also be reviewed.

2. The safety professional should start at the beginning of your process where your raw materials are taken in and go carefully through all your processes, watching each operation and talking to employees, up to the point of shipping out your worksite's product. He or she should:

A. Watch how materials are handled and stored, checking the stability of storage racks and the appropriateness of storage for flammables and explosives.

B. Check the openings that expose moving parts for pinch points and other hazards.

C. Check hand tools and equipment and wiring in the maintenance shop.

D. Arrange to see operations on every shift and to observe any "after hours" operations such as cleanup or fork lift battery recharging.

E. Be interested in how you manage your hazard prevention and control program.

F. Open every door and look in every corner of your facility.

G. Walk around the outside of buildings to check on such things as:

- Chocks for trucks at the loading/unloading docks
- Fork lift ramps
- Outdoor storage of flammables and explosives
- Any fueling areas

H. Suggest jobs for job safety analysis, especially for jobs that might involve ergonomic hazards.

I. Assist you in developing or improving your injury reduction program.

3. The industrial hygienist and occupational health professional should also start at the beginning of your production flow, observe all operations, talk to employees, and follow the production flow to the point of shipping. They will want to:

A. Check your inventory of chemicals against what can be found in the worksite.

B. Know what metals are used in any welding operations.

C. Check any production areas where eating or smoking is allowed.

D. Check for the possible presence of asbestos, lead, carcinogens, and the like.

E. If respirators are used, check if you are using more than one brand, how you fit test each employee, whether pulmonary function testing is done, how the respirators are cleaned, maintained, and stored.

F. Do full-shift sampling of contaminants thought to be present in order to understand the overall exposure to employees.

G. Watch the movements workers make to perform their jobs to see if there are actual or potential cumulative trauma disorder (CTD) or other ergonomic hazards.

H. Possibly suggest processes for routine process hazard analysis.

I. Help you set up or improve regular monitoring programs for any contaminants or other health hazards found to be present.

Note: The items above are listed as "signs" of a thorough survey and do not constitute an exhaustive list of activities to expect.

Follow-up Surveys

You need periodic surveys to take advantage of the rapidly growing scientific and engineering knowledge about hazards and their prevention or control. These surveys also help pick up the hazards that occur as processes and procedures *evolve* over a period of time. The frequency of these surveys will depend upon the size and complexity of your operations.

The baseline survey should provide the basic inventory of hazards and potential hazards of your worksite. The inventory will be improved by what you learn from later periodic surveys, change analysis, and routine hazard analysis, but it is founded on the results of the baseline comprehensive survey. It is very important that this initial survey be done well.

Change Analysis

Before making changes in the worksite, analyze the changes to identify potential hazards they may create. Anytime you bring something new into your worksite, whether it is an entirely new building, a piece of equipment, different materials, or a new process, you could be adding many new hazards. If you are considering a change for your worksite, it should be thoroughly analyzed before it is introduced. Not only is this type of change analysis cost-effective for the human suffering and financial loss it may prevent, but it is also usually less expensive to prevent problems before introducing something new than it is to attempt to fix them later.

The nearest state or federal OSHA office will look at plans, blueprints, and photographs and provide advice as to health and safety concerns. This assistance will *not* trigger an enforcement visit. OSHA enforcement personnel cannot come to the site to see what you are planning; you must bring information to them. OSHA-funded state consultants can come to your site.

Building or Leasing a New Facility

Even something as basic as a new facility needs to be carefully reviewed to identify hazards it might pose to those who will work in it. What may appear delightful to the architect and even seem to enhance production of your product may be a fatal or otherwise harmful management decision that can be avoided by having safety and health experts take a careful look at all the plans.

When you are leasing a facility that was built for a different purpose at a different time, the chance of acquiring new health and safety problems is even greater. A thorough review of the actual facility, plus the blueprints or plans for any renovations, should be made. The most obvious concern about acquiring an existing facility is whether asbestos insulation exists and whether it is friable (flakes off in tiny particles). You may find that something as easy to fix as a loose stair railing has gone unnoticed in the hurry to renovate production areas.

Having expert safety and health professionals involved in the planning of any new or contemplated lease or purchase of an existing facility can save frustration, money, and lives.

Installing New Equipment

Manufacturers cannot always be relied upon to have completely analyzed and prepared controls or safe procedures for equipment as it will be used at your worksite. If the equipment is produced in a foreign nation, it may not even meet clear requirements of U.S. standards and laws. Health and safety professionals should be involved in the purchase decision and in the plans for installation.

Many companies also provide for a period to test newly installed equipment, using their most experienced operators to watch for hidden hazards in the operations before full production begins. As with new facilities, the sooner flaws are detected, the easier and cheaper corrections will be.

Using New Materials

Before introducing new materials to your production processes, research the hazards that the materials themselves present, plus any hazards that may appear due to the processes you plan to use with the materials at your worksite. The place to start is usually with the manufacturer's Material Safety Data Sheet (MSDS). An MSDS is required for all materials containing hazardous ingredients and should arrive with each shipment. It should provide the information that an industrial hygienist needs to analyze the hazard an ingredient presents and how to prevent or control it.

Some traditionally used materials, such as lead in paint, are dangerous to use, but can usually be replaced with less hazardous mixtures. In other cases, you may need to establish controls for the hazards that the material involves.

Starting up New Processes

New processes, even when used with familiar materials, equipment, and facilities, may present new hazards because workers are interacting differently. It is important that safe work procedures for new processes be developed thoughtfully. After the operators have become familiar with those procedures, perform routine hazard analyses (discussed below) to discover any hidden hazards.

Analyzing Multiple Changes

Often several smaller changes will result in one big change. When you start up production of a new product, chances are you will have new equipment, materials, and processes to monitor. Make sure each new addition is not only analyzed alone but also in relation to the other changes.

The information gathered from the analysis of change

at your worksite will be added to your basic inventory of hazards that is the base from which you design your hazard prevention and control program.

PROCESS HAZARD ANALYSIS

For the purpose of this analysis, a "process" can be defined as any series of actions or operations that convert raw material or materials into a product or products. The process can terminate in a finished product ready for consumption, or in production of a product that is the raw material for subsequent processes. A "process hazards analysis" is a detailed study of a process to identify every possible hazard presented to employees.

Every element of the process must be studied. Each action of every piece of equipment, each substance present, and every move made by an employee must be considered to present a hazard to employees until shown by the process hazards analysis either not to be a hazard or to be a hazard that is controlled in every foreseeable circumstance.

Often the process hazards analysis will concentrate on the specialized equipment used in the process to bring about the desired changes. The equipment may be used to:

- Move materials
- Apply mechanical forces to change materials
- Apply special forms or concentrations of energy to materials (e.g., ionizing radiation, magnetic or electric fields, or lasers)
- Mix materials
- Bring and hold materials together and contain them, either under ambient or special conditions, for chemical or biological reactions, and so on

Processes may be either batch or continuous. Some operations may be conducted remotely. There may be special instrumentation to monitor conditions at various stages in the process. These instruments will keep the operator informed and perhaps also directly control the process or shut down operations if a hazardous or otherwise undesirable condition is being approached.

For the most effective protection of your employees (and your investment), the first process hazards analysis should be conducted as a change analysis, when the process is being laid out on paper and before the process equipment is selected. Performed as part of your change analysis, this process hazards analysis will result in a program for hazard prevention and control, including the establishment of an inspection and maintenance schedule for all equipment, including the instrumentation. As discussed above in the section on change analysis, this will minimize the chances for failure or unexpected changes in the process and in monitoring equipment that could result in unexpected hazards. Remember to take into account any substances or mixtures of substances that may present fire or explosion hazards.

You will choose the process equipment not only for its suitability for carrying out the process effectively and efficiently, but also for its capacity for confining the process within predetermined safe limits. The type, number, and location of detectors you select for monitoring the process must not only ensure productive operation but also safe operation.

Remember that even when a process appears initially to be so simple that hazards analysis during the design phase *seems* unnecessary, it should be done anyway. If the process really is simple and there are no potential hazards, then the process hazards analysis will likewise be simple and will take very little time.

If, on the other hand, things are not as simple as they seemed and the analysis reveals potential problems that would have otherwise been overlooked, correction at this early stage will save time, money, and possible injuries and grief. Even when a process hazards analysis is done during the design phase, unforseen problems sometimes appear when operations begin. Once you begin production, it is helpful to review the pre-operational analysis to make sure that you anticipated all hazards, or if there was no prior analysis, to conduct it at that time.

Of course, whenever there is a change in the process or a problem is first noted in an existing process, a new process hazards analysis should be conducted as discussed above under change analysis. As with the periodic comprehensive surveys, process hazards analysis should be done periodically. This is especially true for operations with many complex potential hazards. Companies using hazardous chemicals, explosives, or large quantities of flammable liquids or gasses should do continuous process hazards analysis, moving through their various processes at least every five years. These reviews may be called by a different name, such as "process review" or "hazardous operations review."

Process hazards analysis will include hazards such as those associated with:

- Mechanical and chemical operations
- Low and high temperature and pressure operations
- Possible high levels of radiant energy
- Direct contamination of employees
- Contamination of the air with toxic substances

Remember that toxic substances may be the raw materials entering the process, the intermediate products, the final products, or the by-products or waste products.

Who Should Do the Process Hazards Analysis

For processes involving high-hazard chemicals, explosives, and/or flammable liquids or gasses in large quantities, process hazards analysis should be conducted by a team rather than just one person. At least one member of that team should be knowledgeable in engineering and process operations, and at least one other should be intimately familiar with the process being analyzed. Small businesses may need to hire expert consultants to help with the analysis. If you have a small business and you do not know whether your process is one that needs a high level of expertise, you can call your local office of the OSHA-funded, state-run consultation service or the nearest state or federal OSHA office for advice.

For less hazardous operations, the analysis can be done by someone who is familiar with the operation following the steps outlined here.

Overview of the Process

Start by writing down the answers to these questions:

1. What kind of process is it? Mechanical? Chemical? Biological?
2. What is the product?
3. What is the rate of production?
4. What raw materials will be used? How much?
5. Will there be intermediate products? What will their quantities be?
6. Will there be waste materials that will be a problem because of their toxicity and/or their quantity?
7. Are there inherent hazards in this process that indicate that you should look for a safer way to produce the product?

8. What kinds of equipment are used in the process? Is other equipment available that would be safer?
9. Is there sufficient and reliable monitoring and control equipment? Is it fail-safe in all instances?
10. What are the various workers' roles?
11. Where they work directly with substances and equipment, are workers' operations as safe as possible?
12. Are there points in the process where workers' exposure to hazards could be reduced?
13. Could any emergency situations develop? How many unexpected events could happen at the same time and what would be the result?

With the answers to these and other pertinent questions, prepare a narrative report in some detail, which would be intelligible even to persons unfamiliar with the process.

Process Flow Chart

Unless your process is very simple, with only one or two steps, you should prepare a diagram or flow chart of the process. The flow chart is a detailed expansion of the overview and is prepared as a visual and verbal description of each step in the process, relating each step to the others. The flow chart helps provide order to the analysis by laying all elements of the process before you in a clear and related way. A careful review of the flow chart with reference to the process should reveal any elements that were overlooked in the overview. An example of a Process Hazard Analysis Worksheet is shown in Appendix 4.

Hazardous Substances

Look at each substance in the process:

- Raw materials
- Intermediates
- Final product
- Byproducts
- Wastes

Be sure to look at all activity involving substances that could produce hazardous conditions or products, or that could lower the oxygen content of the air in any inhabited space, particularly in work involving confined spaces. Study the potential hazards of each substance—toxicity, fire, or explosive hazard and others (a requirement of 29 CFR 1910.1200[d]). See how each substance appears in the process and in what quantity. Is each hazardous substance being handled so as to minimize its hazard? Could a less hazardous substitute be used? With these answers, you can plan your program of prevention and control. Take the precautions necessary to protect your workers, beginning with substitutions and engineering controls. Establish work practices and utilize personal protective equipment as necessary. (Hazard prevention and control is covered in Chapter 10.)

Equipment

Look at all the equipment, from the equipment used to move the raw materials in at the beginning of the process to the equipment used to move the products, by-products, and waste products out at the end of the process. It must be safe not only for the operator but also for other workers in the vicinity. Look at the materials handling equipment throughout the process. You will want to examine all the machines used to bend, form, cut, grind, mill, smooth, or otherwise change the surface or contours of solid materials. Also look at those that weld, crimp, rivet, or otherwise fasten one solid piece to another.

For liquids and gasses, there will be enclosed systems for transfer and storage, vessels for mixing, and reactor vessels that may be subjected to high or low temperatures and/or pressures. If there are exothermic reactions—reactions that release heat—are there adequate provisions for removing the heat? Look at any special equipment that may produce hazardous levels of radiant energy. Once you have determined the potential hazards, you can plan your prevention and control by using the safest equipment, implementing engineering controls and work practices, and providing and ensuring the use of personal protective equipment as necessary.

Worker Exposures

Look at each worker's locations and actions relative to all hazards throughout the workday to determine all interactions with substances and equipment. Do certain actions of a particular worker result in a hazard for that worker or others? Employee exposures to physical agents such as microwave radiation will have to be measured with special instrumentation. Air contaminants will have to be measured to determine exposure levels. When outside doors and windows are closed in cooler weather, will air contaminant levels and exposure to workers rise? Once you know the exposures, you can plan your prevention and control, utilizing engineering controls, work practices, and personal protective equipment as necessary.

Preparing for the "Unplanned Event"

Especially when you are dealing with high-hazard chemicals or volatile explosives, you will need to go beyond analysis of hazard potential during times when the process works as expected. Using analytical tools such as "what if" or "fault-tree" analysis, you can determine most of the possible process breakdowns and then design prevention and control for the possible cause(s).

"What if?" analysis starts with points in the process where something could go wrong and determines what else could happen and what all possible outcomes would be. For those potential unplanned events that could possibly contribute to an undesirable outcome, additional prevention and controls must be planned.

"Fault-tree" analysis works in the other direction. It starts with any undesirable outcome that might be possible (even though very unlikely) given the potential hazards involved in various processes. Then it traces back into the processes to identify the events or combination of events that would have to occur to produce that outcome. Once you have established what those events are, you can design prevention and control to make it impossible or virtually impossible for them to occur.

There is considerable literature on both of these types of analyses that can provide step-by-step instructions on how to do them. This type of analysis is really meant for high-hazard operations. If you have a small business and are not sure if your processes should be analyzed in this manner, you can ask for advice from your OSHA-funded, state-run consultation service.

As with the other hazard analyses and surveys discussed here, process analysis should be repeated at intervals suitable to the type, complexity, and hazardousness of the process. If the process falls into the category described above of high-hazard chemicals, explosives, or flammable liquids or gasses in large quantities, the analysis of the process should be repeated at least every five years.

Activity Hazard Analysis

In construction and other industries involving a rapidly changing work environment, different contractors, and widely different operations, phase hazard analysis is a helpful tool. Activity is defined as an operation involving a type of work presenting hazards not experienced in previous operations, or where a new subcontractor or work crew is to perform work. In this type of hazard analysis, prior to beginning each major phase of work, the contractor or site manager should assess the hazards in this new phase, not only to coordinate appropriate supplies and support but also to prepare for the hazards that can be expected and prepare a plan to eliminate or control them.

To find these hazards and to eliminate or control them, you will use many of the same techniques that you use in routine hazard analysis, change analysis, process analysis, and job analysis. One major additional task will be to find those hazards that occur when combinations of activities occur in close proximity. Workers for several different contractors of different expertise may be intermingled and need to learn how to protect themselves from the hazards associated with the work of nearby colleagues as well as with their own work.

Job Safety Analysis

Job safety analysis, the most basic and widely used routine hazard analysis tool, consists of breaking a job down into tasks and each task into the steps involved. Each step is then analyzed for potential hazards, then prevention or control of these hazards is designed. As a result of job safety analyses, new safe work procedures can be drawn up and workers taught to perform them. Many companies successfully use operators who do the jobs as part of the teams that perform the analysis.

Note: It is best to keep the written analysis of a job separate from the description of procedures that workers follow to accomplish the job safely. A good job safety analysis will involve a level of detail that would be cumbersome to use for job procedures. When a written analysis is planned to be used to describe job procedures, it tends to become too superficial.

Involving Workers in Establishing the Inventory of Hazards

Whenever possible, you should use the special knowledge that your workers possess from their intimate involvement with equipment, materials and processes. Workers should be encouraged to have open access to the professionals who do the comprehensive surveys. Workers are frequently used in change analysis of new equipment and/or processes for their valuable insights on how things will "really work." As mentioned above, many companies regularly include hourly operators in job hazard analysis, and they can play a similar role in process hazards analysis. In addition to the benefit that you receive from their insights, they also benefit by better understanding hazards and their preventions and controls. This helps them do a better job of protecting themselves and their fellow workers.

Using the Inventory of Hazards

As previously mentioned, the surveys and analyses described here provide an inventory to use in planning a program of hazard prevention and control (explained more fully in Chapter 10). Briefly, prevention by substitution with less hazardous materials or equipment should occur when possible. Engineering controls to remove the worker from the hazard are the next best. For the rest of the hazards, safe work practices, in which workers have been adequately trained and which are consistently enforced, should be designed along with any personal protective equipment needed. In some rare instances, administrative controls such as worker rotation or more frequent work breaks will also be needed.

Summary

Establishing a complete hazard inventory is not as complicated as it may sound. It begins with having industrial hygiene, safety, and occupational health experts survey your worksite to determine just what your hazards and potential hazards are. Periodic surveys (done at intervals that make sense for the size and complexity of your worksite) use any new engineering or scientific knowledge of hazards and their prevention. They can also help find new hazards that have evolved with work procedures over time.

Change analysis simply means making sure that individuals who are knowledgeable in worker health and safety help design and plan for changes in your worksite to prevent expensive problems before they occur. Change analysis uses elements of routine hazard analysis depending upon the type of change.

Routine hazard analysis also adds to your inventory of hazards and enables you to control hazards that develop in work procedures or within processes, or that occur because of the changes in the workplace due to varied phases. Many of these actions and tools overlap to a certain extent to ensure total coverage.

For small businesses, the OSHA-funded, state-run consultation services can help with the creation and updating of the hazard inventory. Larger businesses may wish to hire private consultants or call upon the assistance of company or corporate experts. Businesses of any size may ask for expert advice from state and federal OSHA offices without fear of enforcement action, but they will have to send or take the information to OSHA enforcement personnel.

Remember, making sure that you know what hazards need to be prevented or controlled is the first step in protecting your workers.

9

Catching the Hazards That Escape Controls

Ideally, each employer should know exactly the hazards and potential hazards to which his or her employees could be exposed and has designed a perfect system of prevention and control of those exposures. In reality, some hazards may have escaped detection to begin with. Other hazards have a way of slipping out of the controls that are set up to protect workers. So you need ways to catch these hazards and get them controlled, or better controlled, before anyone gets hurt or before too many workers get hurt.

As a first step, a complete worksite analysis should be conducted to provide the basis for hazard prevention and control. This analysis includes identifying all hazards and potential hazards through comprehensive surveys, change analysis, and routine hazard analysis; regular site inspections; employee reports of hazards; accident and near-miss investigations; and analysis of injury and illness trends over time.

This chapter will cover regular site inspections; employee reports of hazards; accident and near-miss investigations; and analysis of trends of illnesses and injury. These four tools help to fill in any possible gaps left by the tools described in Chapter 8. At the same time, they help to quickly discover and allow correction of hazards for which the controls have failed, at least temporarily. These are also the tools that provide information to improve the "foolproofing" needed to keep human error from causing or allowing those controls to fail.

Regular Site Inspections

Inspections are the best understood and most frequently used tool of worksite analysis. Much has been written about them and many check lists are available in other publications. Check lists should be developed to address the unique conditions of a specific company, facility, operation, or process. Although there are many publications available containing generic check lists, these will have to be modified to meet the specific requirements of the check list location.

In this discussion, we will concentrate on some of the things concerning inspections that do not appear frequently in other publications.

Definitions

The term *inspection* is used to cover many kinds of actions that involve looking closely at something to see if it meets requirements. Several kinds of inspections are probably done at your worksite, many of them at fixed intervals. For the purposes of this chapter, the term *regular site inspection* means a general inspection of every part of the worksite to locate any worker hazards that need correction. The term also includes routine industrial hygiene monitoring and sampling.

Inspection Frequency

This general worksite inspection is done at specified intervals. It is usually recommended that medium and large fixed worksites be completely inspected at least every quarter, with some part of the inspection occurring each month. (For construction sites, it is recommended that weekly site inspections be conducted because of the rapidly changing nature of the site and its hazards.)

Smaller fixed worksites might not need to have general inspections every month, but their inspections should cover the whole site at one time. Even for the smallest worksite, inspections should be done at least each quarter. If the small worksite uses hazardous materials or involves hazardous procedures or conditions that change frequently, inspections should be done more often. Employers in small businesses may contact the local office of the OSHA-funded, state-run consultation service for assistance in determining the most appropriate frequency of inspections.

What Should Be Inspected?

Inspections should follow check lists or written guidance based on the inventory of hazards you have established following the steps described in Chapter 7, and the preventive actions and controls that were designed to reduce or eliminate worker exposure. Regular site inspections should be designed to check each one of those controls to make sure that hazards are contained. Some suggested broad areas of inspection are listed in Appendix 5.

Hazards may be controlled by engineering. Examples of engineering controls include guards placed on equipment to prevent hand or body contact with tools or machinery that can cut or pinch body parts or catch clothing, hair, or body parts and pull them into the machinery. These guards should be checked during inspections to make sure that they are in good working order and that they remain in place during operations. Another example of engineering control of hazards is ventilation systems to carry away air contaminants. The velocity of the ventilation system should be checked periodically and, depending upon the risk of the contaminant controlled, periodic air sampling should be done as well.

Hazards may also be controlled by specifically designed work practices. An example is provided by the precautions taken by the operator of production equipment when restarting after a "jam." The inspector should keep an eye out for any unauthorized modifications of those work practices.

Another way to control hazards is through personal protective equipment (PPE). Meat cutters, for example, should wear metal mesh gloves. Many varied types of worksites require safety glasses or hearing protection in certain areas of operation. The inspector should look carefully both at the equipment itself to ensure good condition and at personnel in the areas where PPE is required to make sure PPE is being worn and worn appropriately. This includes stopping workers who have placed disposable hearing protection only part-way into their ears, leaving most of the "plug" protruding. The inspector will insist that the PPE be worn as intended, to reduce risk.

Inspectors should make sure that they are looking at areas outside of the mainstream of production. These areas should be included in the check list.

Who Should Inspect?

Ideally, medium and large worksites will have more than one type of regular site inspection.

Supervisors

Many places make it the supervisor's responsibility to inspect his or her work area at the beginning of every shift to ensure that equipment and personnel are ready to work safely. This can be particularly helpful when other shifts use the same area and equipment or when maintenance and/or cleaning is done after hours. However, supervisors' inspections of their own areas should not substitute for the broad general inspection recommended in the these guidelines. There are two reasons for this:

1. Fresh eyes are needed for the general inspection. Those who work in an area can start "not seeing" things that they get used to. It is always good to have cross-inspections where supervisors or employees from one area will be looking at other areas and vice versa.
2. The general site inspection should be broader than just specific work areas where individual supervisors have responsibility. Outdoor and other common areas should also be covered in the general inspection.

Employees

We recommend involving employees in the safety and health program, in both problem identification and resolution. One of the ways used frequently to involve employees in the safety and health program is an employee committee or a joint management-employee committee to do routine inspections. This serves two purposes:

1. It expands the number of people doing inspections and, therefore, the number of eyes looking for hazards.
2. It increases employee awareness of the safety and health program.

Safety and Health Staff

The most common—and the most logical—inspectors are staff personnel who specialize in safety and health. Even where other inspectors are used, it is wise also to involve the specialists. In a small business, the "specialist" may be the Director of Personnel or another member of management with many important duties besides safety and health. Using the person(s) responsible for safety and health in the worksite serves two additional functions:

1. It keeps the person responsible for safety and health in touch with the success and/or problems with the hazard prevention and control program.
2. It uses your greatest in-house source of expertise.

What Training Should Inspectors Have?

Safety and Health Staff

Personnel with responsibility for the development of the safety and health program should have at least the equivalent of the OSHA Training Institute course number 501, "A Guide to Voluntary Compliance in Safety and Health." Larger worksites and small worksites with hazardous operations or materials should have more.

For advice about training and information about available OSHA training, write or call the OSHA Training Institute:

OSHA Training Institute
1555 Times Drive
Des Plaines, Illinois 60018
(708) 297-4913

Cost for private-sector course participants is minimal and is based on the number of days of the training. In 1990, course costs ran between $60 for a two-day course to $535 for a fourteen-day course. In 1990, number 501 cost $173 as a five-day course.

Supervisors

All supervisors should have training on the hazards that workers under their supervision are likely to encounter, and training on how to control them. When they are responsible for area inspections, they should also have specific training on how to inspect. This does not necessarily have to be formal training, but it should be provided by someone who is knowledgeable.

Employees

All employees should understand the potential hazards that they might be exposed to and how to protect themselves and their fellow workers. Those who are involved in inspections need training on all the potential hazards of the worksite and how they are controlled. They will also need written guidance, tips for inspecting, and some on-the-job training by the safety and health staffs or "specialists."

Written Inspection Reports

In all but the smallest and least dangerous of sites, written inspection reports are necessary to record the hazards discovered, the responsibility assigned for correction, and the tracking of correction to completion.

Having a written record will help ensure:

- Assignment of responsibility for hazard correction
- Tracking of correction to completion
- Identification of more complex problems with controls when the same type of hazards keep popping up even after correction is verified
- Identification of problems with accountability systems for those who are charged with keeping hazards controlled and/or with ensuring hazard correction
- Identification of hazards for which no prevention or control has been planned

Of course, having such written records will help the most if they are read by someone knowledgeable in the safety and health program who can provide summaries of problems to top managers.

Tracking Corrections of Hazards

Tracking of hazard correction is covered in more detail in Chapter 10. It is important to remember, however, that the quality of follow-up to inspections can make them successful or make them failures. If correction cannot be accomplished immediately at the time of discovery, the inspection report should not be considered closed out until the *final* correction has been made. A written tracking system will help make the inspection program succeed. The best tracking system is right on the inspection report form itself.

Information obtained from regular general inspections should be helpful in updating or adding to your inventory of hazards and/or making improvements to your program of hazard prevention and control.

The important things to remember about routine general inspections are:

1. They should cover every part of the worksite.
2. They should be done at regular intervals with frequency depending upon the size of the worksite and the nature of the hazards.
3. "Inspectors" should be trained to recognize hazards and be able to bring "fresh eyes" to work areas inspected.
4. Hazards found must be tracked to correction.
5. Information from inspections should be used to add to the inventory of hazards and/or improve the hazard prevention and control program.

Employee Reports of Hazards

Employees play a key role in helping you discover and control the hazards that may occur, or may already exist, in your workplace. They have a unique perspective on procedures and conditions since they are the ones who are closely involved with them.

A reliable system for employees to report these conditions is an important aspect of an effective safety and health program. To be effective your system will need the following characteristics:

- A genuine company or worksite policy, which is consistent with other worksite policies, to encourage employees to report their concerns about

safety and health conditions or possible hazards in work practices
- Timely and appropriate responses to the reporting employee
- Timely and appropriate action where valid concern exists
- Tracking of any hazard correction required
- Protection of reporting employees from official and unofficial harassment

Company Policy

When you have decided what the policy will be, it is important that all employees understand what the policy is and that it is "real." In larger worksites, the policy should be typed and placed on bulletin boards, distributed to all employees, and discussed in weekly or monthly safety meetings. In the smallest worksites, gathering everyone together to go over the policy and inviting discussion or questions may be sufficient. You will know that you have done enough when every employee, when asked, can tell you what the policy is.

The written policy is a good place to affirm your intention that reporting of hazards will be protected from harassment or reprisal of any kind.

Timely and Appropriate Response and Action

A system of having employees report hazards will not work if they cannot see reasonable results in a reasonable period of time. A preliminary response can be given where time is needed to analyze the reported hazard. The employee should be kept informed. Many larger workplaces assign special maintenance codes to work orders involving safety or health. This requires the maintenance supervisor to give that work higher priority. When complete correction of a hazard requires ordering parts or materials that take several months to arrive, the employee should be given a status update from time to time so that he or she will know that the concern has not been forgotten.

Just as the results must be timely, they must also alleviate the employee's concern. If management decides that the reported "hazard" is not really a hazard, the reason why this is so should be carefully and completely explained to the reporting employee. Care should be taken to express gratitude for "erring" on the right side; that is, the employee should be thanked for bringing it up. It is better to have some nonhazards reported than to miss even one real hazard because an employee believed that no one would take it seriously.

If management is not sure whether the reported practice or condition is really a hazard, further checking should be done. If a good description can be given over the phone, or photographs or drawings show it well, management can call or visit the nearest federal or state OSHA office for advice. Small businesses may contact the nearest office of the OSHA-funded, state-run Consultation Program.

If the employee has discovered a real hazard, the action taken by management to correct it should be appropriate and should be completed as soon as possible.

Hazard Correction Tracking

Each valid hazard identified through employee reports of hazards should be tracked to complete correction. Hazards that are quickly corrected may not present a problem for long-term tracking, but a record of the correction will help determine where management systems have broken down if the same hazard is found again. Different management measures may be needed for hazards that do not *get* corrected from those that do not *stay* corrected. For hazards that require complicated or time-consuming correction, a system of tracking is needed to make sure that the final correction is not forgotten in the press of other matters. Additionally, tracking of long-term hazard correction will enable management to keep the reporting employee better informed.

Protection from Harassment

It is important that employees feel secure that reporting a hazard will not result in any official or unofficial harassment or reprisal from management, individual supervisors, or co-workers. The policy on employee reports of hazards should make that clear. In addition, there are several ways to help ensure that harassment is never considered:

1. Avoid rating supervisors negatively for employee reports of hazards in their areas as long as they are reacting appropriately to the reports.
2. Disassociate employee reports of hazards and the disciplinary system. Even placing them together on a bulletin board or in sequence in an employee handbook can give the impression that one employee can get another in trouble by reporting him or her for hazardous practices.
3. Approach all discussions and written descriptions of the employee reports of hazards system as a group effort to keep the worksite safe and healthful. Emphasize the positive.
4. Emphasize the responsibility that each employee has

for his or her fellow workers' safety and health as well as for his or her own. The safety and health of any employee in the plant should never be considered "none of your business."

5. If you find a case of harassment for reporting a hazard, enforce your policy clearly and emphatically.

Reporting Systems

There are several kinds of systems to use for employee reports of hazards. Many larger worksites will use a combination of some or all of these systems. The most common systems are:

- Oral reports to supervisors
- Suggestion programs
- A "stop" card program
- Maintenance work orders
- Written forms providing for anonymity

Oral reports

Every worksite should be able to use oral reports of employees to their supervisors. If the supervisor is properly trained and knows his or her responsibility for the safety and health of the workers under supervision, informal reports made orally are very natural. In the case of valid concerns, the supervisor then has the responsibility either to correct the hazard, request correction by maintenance, or ask for assistance from the safety department.

Most worksites encourage this type of reporting. Used alone, however, it does not provide for hazard correction tracking or the ability to look for trends and patterns. It also provides little protection from supervisors who may not be as concerned about health and safety as they should be.

At very small worksites, where "everybody knows everything," this oral system may be all that is needed. We recommend, however, at least a system for the supervisor to make a short report of each hazard reported and the action taken. Worksheets in Appendix 6 shows a format that could be used for such a report.

Suggestion Programs

The most frequently used type of *written* system is a suggestion program where employees are encouraged to make safety and health suggestions. This has the advantage of being a very positive approach and encourages the use of imagination to suggest safer or more healthful ways of doing things as well to report unsafe conditions and acts. If such programs are being used for safety and health reports of hazards, however, management must ensure that collection points are checked several times a day and that suggestions are read at the time of collection to ensure that identified hazards get corrected in a timely manner.

If the suggestion program is used as the sole means of reporting hazards or the sole means of reporting them in writing, management must also work hard to ensure that employees understand that they are encouraged to use the system for all types of safety and health hazard reporting and not just for ideas in the class of "it would be nicer if . . ."

"Stop" Card Program

Many medium and large worksites develop or buy and use a program for employee reports of hazards such as the one developed and sold by DuPont. This program includes the format for training employees in basic hazard recognition. Employees jot down information about unsafe conditions and practices on special cards, which are usually turned in to the safety department for checking and tracking of any valid hazard correction.

Some workplaces give awards for the highest number of cards with valid concerns turned in over a specific period of time. Others have set quotas for number of cards turned in. The success of these particular uses seems to depend upon the "culture" of the worksite.

Maintenance Work Orders

For unsafe conditions, usually maintenance will have to be called to make the correction. Some companies give every employee the right to fill out a maintenance work order; others will allow employees to fill them out but require supervisory sign-off before they are sent to the maintenance department.

Note: This system should only be used by employees reporting hazards if there is a special high-priority code for maintenance work orders involving safety or health, so that the maintenance supervisor is required to give them higher priority than maintenance for production improvement only. Copies should also be carried immediately to the Safety Department (or the person in charge of safety and/or health) so that correction can be tracked.

This special code is also helpful in allowing the safety department or in-house safety and/or health expert to check over time for patterns that may indicate a need for a closer look at conditions or practices.

The work order system is not sufficient when used by itself because, although it can lead to correction of hazardous conditions, it cannot correct hazardous practices. Also, the work order system alone does not provide for more imaginative approaches to standard conditions or procedures to make them safer or more healthful.

Written Forms

While some of the systems described above include filling out forms, the best written system for your worksite may be one that you devise especially for employee reports of hazards. You can allow for anonymity, when desired, by not requiring the reporting employee to sign his or her name or to give the filled-out form directly to a supervisor. Anonymous reports can be addressed by posting a typed response on a bulletin board in the area mentioned in the report.

Remember that in all of these aspects discussed above, some variations will work better for your site than others. Just make sure you have a policy that encourages employee reports of hazards, make this policy well known and understood, protect reporting employees from harassment, respond appropriately in a timely manner, and track all hazards to correction. Use the information you obtain about hazards to revise your hazard inventory and/or improve your hazard prevention program.

Accident/Incident Investigation

Much has been written about investigating accidents and many elaborate charts have been devised to assist the investigator. In this chapter, we will not attempt to recreate all the information readily available elsewhere. (A good, comprehensive publication on accident investigation available from the National Safety Council is "Accident Investigation: A New Approach," published in 1983.)

Accident/incident investigation is another tool to use in discovering hazards that either were missed when the hazard inventory was set and used to develop the hazard prevention and control program or have managed to slip out of the controls planned for them. But it is *only* useful as a tool when it is done with the objective of discovering *every* contributing factor to the accident/incident in order to "foolproof" the condition and/or activity to the extent feasible to prevent future occurrence. In other words, you want to identify the root cause of the incident or accident.

This section defines the words *accident* and *incident*, and discusses what should be investigated, who should investigate, investigation training, results desired, and the use of those results in updating or revising the hazard inventory and/or improving the hazard prevention and control program.

Definitions

The National Safety Council defines an accident as "an unplanned, undesired event that results in personal injury or property damage," and defines an incident as "an unplanned, undesired event that adversely affects completion of a task." You will notice that incidents, therefore, include accidents. Note also that accidents include incidents involving property damage but no personal injury. "Personal injury" is used to cover illness that results from a single occurrence but not occupational illness that results from long-term exposure to health hazards.

What Should Be Investigated?

Since any incident (unplanned and undesired event) could result in property damage or personal injury, it should be investigated to determine the contributing causes and actions needed to prevent future occurrences. That includes near-misses where no property was damaged and no personal injury was sustained but where, given a slight difference in time or position, damage and/or injury could have occurred.

Who Should Investigate?

The usual investigator for all incidents is the supervisor in charge of the area and/or activity involved. The Safety Department or the person in charge of safety and health should at least review those investigations and provide for another level of investigation when:

1. The incident had very serious results.
2. The nature of the incident is very complex.
3. The incident involves more than one supervisor's responsibilities.
4. The initial investigation has not clearly established a full range of contributing factors and/or preventive actions.

Many companies provide for a team (or a subcommittee of the joint employee-management committee) to investigate incidents involving serious injury or extensive property damage. This may supplant the supervisor's investigation or may be done as a second-level investigation. When the team or committee investigation is used, the team leader or chairman must have enough authority and status in the organization to get done whatever is needed.

Incident Investigation Training

No one should investigate incidents without appropriate accident investigation training. This type of training is easily available from many consultant and safety and health professional organizations. To be sure the training is going to be on track, you can check its contents against the information found in the National Safety Council's pamphlet "Accident Investigation: A New Approach." Training should be followed by checking investigative reports to see if the training is being used appropriately.

Results Desired

No matter what its length or style of presentation, the investigation report should carefully document the full range of facts. It should make clear that everyone with any knowledge of the incident was interviewed thoroughly, using the six key questions of who, what, when, where, why, and how. Fact and opinion should be presented separately and clearly as what they are. A good investigation will result in several contributing factors being named and several preventive actions being recommended.

One trap specifically to be avoided is that of laying the sole blame on the injured employee. Even if the injured employee immediately blames himself or herself for "making a mistake" or not following the precise prescribed procedure, the accident investigator must not be satisfied that all contributing causes have been identified. The error made by the employee may not even be the most important contributing cause. The employee who has not followed the prescribed procedure may have been directly or indirectly encouraged by a supervisor to "cut corners." Or the prescribed procedure may not be practical, or even safe, in the eyes of the employee. Sometimes the elaborate procedures required are difficult to follow correctly and engineering redesign may be a better answer. In such cases, other management errors may be the most important contributing causes.

When reviewing accident investigation reports, the safety department or in-house safety expert should be on the lookout for catch phrase "causes," such as "employee did not plan job properly." This type of catch phrase may have a meaning "between the lines," but it is not conducive to identifying all possible causes and all possible preventions and controls. Certainly it is too late to "plan for a job" when the employee is about to do it. Further, it is unlikely that safe work is accomplished when each employee is encouraged, let alone expected, to plan procedures alone. All supervisors and others who investigate incidents should be held accountable for being clear in their descriptions of causes.

The approach to preventive action recommendations should be to list all the ways the condition or activity can be "foolproofed," with an eye to making it very difficult (or close to impossible) for the incident to happen again. Cost or engineering concerns should not enter in at this point. Top management should have the benefit of complete thinking before any decisions are made about preventive actions. Some actions will be such that they can be accomplished immediately; others may take time, planning, and capital expenditure.

Use of Accident Investigations

Beyond the immediate use of accident investigations to prevent future occurrences, the information obtained through the investigation should be used to update and revise the inventory of hazards and/or the program for hazard prevention and control.

Analysis of Patterns

The Safety and Health Management Guidelines list the analysis of "injury and illness trends over time, so that patterns with common causes can be identified and prevented," as the last action under Worksite Analysis. The review of OSHA illness and injury logs is the most common form of pattern analysis. It is not, however, the only useful pattern analysis. Pattern analysis may also be done with any record of hazards. Inspection and employee hazard notification records are examples.

Pattern Analysis of the OSHA Log of Injury and Illness

Period of Time Covered

To be useful, any record to be analyzed for patterns must have enough entries so that any patterns that may exist have a chance to form within the record being analyzed. A worksite that has few employees or does very little hazardous work may need to review three to five years of records to allow patterns to emerge. Just because a site is small or does little hazardous work does not mean that pattern analysis is not useful. Even if an office operation has only one or two injuries each year, a five-year review may indicate uncontrolled cumulative trauma hazards or lack of attention to tripping hazards. Larger sites will find useful information in yearly, quarterly, or monthly reviews.

What to Look For

Similar types of injuries or illnesses indicate a hazard or type of hazard that has not yet been controlled. Clues for the diagnosis are found in noting where the injury or illness occurred, type of work being done, time of day, similarity of equipment, and so on. Similar injuries need not be the same injury; they may be to the same part of the body but in a different form. Obviously, repeat occurrences of the same injury or illness demonstrate that hazard controls are not working adequately. Any clue that several injuries are connected to each other in a way that has not been noticed previously is worth investigating further.

Pattern Analysis of Inspection and Employee Report of Hazard Records

Identification of hazards should occur with more frequency than incidents/accidents. It should be possible, therefore, to review records covering shorter periods of time and still be able to see patterns emerge. Repeat hazards, just like repeat injuries, mean that controls are not working. The way to make the control successful may be as "simple" as improving accountability. Causes of hazards may be investigated using techniques similar to those developed to determine causes of accidents.

Summary

Even after you have done comprehensive hazard surveys, analyzed each workplace change for hazards, routinely analyzed jobs and/or processes for hazards, and developed a program of hazard prevention and control, there will still be some hazards in your worksite. They may have been missed or the controls may not have been adequate to keep them controlled or prevented.

Routine general inspections, employee reports of hazards, accident/incident investigations, and analyses of the patterns of incidents and hazards will help you learn more about these hazards and how, not only to correct them, but to keep them controlled.

10

Establishing Hazard Prevention and Control Programs

Are *all* the potential hazards in your workplace prevented or controlled? Do you know that you are required to ensure that they are? The Occupational Safety and Health Act of 1970 requires that each employer "furnish to each of his employees employment and a place of employment which are free from recognized hazards that are causing or are likely to cause death or serious physical harm" (29 U.S.C. 651, Sec. 5 [a] [1]). To do this, an employer must know what hazards exist in his or her worksite and how to prevent or to control them.

The Safety and Health Program Management Guidelines urges employers to identify all hazards and potential hazards and to provide a program of hazard prevention and control "so that all current and potential hazards, however detected, are corrected or controlled in a timely manner."

This chapter will deal with the management systems used to prevent and control hazards: the establishment of controls through engineering, work practice, personal protective equipment (PPE) and/or administrative arrangements; systems to track hazard correction; preventive maintenance systems; and emergency preparation and medical programs.

The goal of the hazard prevention and control program is to "foolproof" the workplace and its operations, as far as is feasible, to keep employees from being harmed. As such, the program is never finished. Preventions and controls are designed and implemented and then revised and improved as changes are made in the worksite and as information about hazards is increased.

The most frequent sources for updating hazard information are routine general inspections, employee reports of hazards, and accident/incident investigations. Other good sources for hazard information updates are the ongoing job hazard analyses, process and phase hazard analyses, change analyses, and periodic comprehensive hazard surveys.

Employers in small businesses may contact the nearest office of the OSHA-funded, state-run consultation service for advice and assistance in developing a hazard prevention and control system. Larger employers may receive assistance for specific areas of difficulty over the telephone or by letter from the nearest state or federal OSHA enforcement office.

Selecting the Best Control

Terminology

Hazards take many forms, including air contaminants, tasks involving repetitive motions, equipment with moving parts or openings that can catch body parts or clothing, microorganisms, extreme heat or cold, noise, and toxic liquids. The terms used to describe the principles of engineering control here may sound a little strange when applied to some of these hazards; however, the concepts do fit.

The information presented here has been organized in a way that will be useful for you. You may find that others will use the terms somewhat differently. There should be agreement, however, about the concepts that the terms have been chosen to describe.

Engineering Controls

To the extent feasible, the work environment and the job itself should be designed to reduce or eliminate exposure to hazards. This approach is called engineering control, but it does not necessarily mean that an engineer is required to design the control. Engineering controls can be very simple in some cases. They are based on the following broad principles:

1. If feasible, design the facility, equipment, or process to remove the hazard and/or substitute something that is not hazardous or is less hazardous.
2. If removal is not feasible, enclose the hazard to prevent exposure in normal operations.
3. Where complete enclosure is not feasible, establish barriers or local ventilation to reduce exposure to the hazard in normal operations.

Elimination of Hazards Through Design

Designing facilities, equipment, or processes in such a way that the hazard is no longer present, even potentially, is obviously the best protection to the worker. Some examples of this are:

- Redesigning, changing, or substituting equipment to remove the source of excessive temperatures, noise, or pressure
- Redesigning a process to use less toxic chemicals
- Redesigning a work station to relieve physical stress and remove ergonomic hazards
- Designing general ventilation with sufficient fresh outdoor air to help prevent "sick building syndrome" and generally provide a safe, healthful atmosphere

Enclosure of Hazards

When hazards cannot be removed or replaced with a less hazardous form, the next best control is enclosure. Enclosing a hazard usually means that there is no hazard exposure to workers during normal operations. There would still be potential exposure to workers during maintenance operations or if the enclosure system breaks down. For those situations, additional controls such as work procedures or PPE might be necessary to control exposure.

Examples of these designs include:

- Complete enclosure of moving parts of machinery
- Complete containment of toxic liquids or gasses from the beginning of the process using or producing them to detoxification, safe packing for shipment, or safe disposal of toxic waste products
- Glove-box operations to enclose work with dangerous microorganisms, radioisotopes, or toxic substances
- Complete containment of noise-, heat-, or pressure-producing processes with materials especially designed for those purposes

Barriers or Local Ventilation

When the potential hazard cannot be removed or enclosed, the next best design approach is a barrier to exposure or, in the case of air contaminants, local exhaust ventilation to remove the contaminant from the workplace. Since there is potential exposure to the worker even in normal operations, this type of engineering design control will need other types of controls to be used in conjunction with it, such as specifically designed safe work practices and/or personal protective equipment (PPE). Examples include:

- Ventilation hoods in laboratory work
- Machine guarding, including electronic barriers (unless they are virtually impossible to bypass)
- Isolation of a process in an area away from workers, except for maintenance work
- Baffles used as noise-absorbing barriers
- Nuclear radiation or heat shields

Work Rules and Procedures

Even when the hazard is enclosed, there will still be times when exposure is possible, such as when maintenance is necessary or when the enclosure system suffers a partial or complete breakdown. When the hazard is only partially controlled by a barrier or local ventilation, there is a good possibility of exposure in normal operations. This is where work practices become important, along with PPE in many cases. For the purposes of our discussion, work practices are divided into safe work procedures and general work practices. General safety and health rules are also hazard controls. These will usually be part of overall workplace rules that also include nonsafety and health items.

The safety and health rules that you develop and make part of your workplace rules also help to control hazards. Written rules play a major part in describing acceptable behavior and identifying actions that are not appropriate. Safety and health rules are most effective when they are written, posted, given to all employees affected by them, and discussed with the employees.

Many employers show employees the link between safety and health rules and the consequences of breaking them by going over the list of rules with the employees. They then ask the employees to sign a statement at the bottom of the list showing that they have read the rules and understand them, and that the consequences of breaking the rules have been explained to them. Both the employer and the employees keep a copy of this signed statement.

Some employers ask their employees to help develop and then revise as necessary the workplace rules. If employees help decide what the rules should be, they are more likely to understand and follow them.

In developing workplace rules you may find it useful to make a list of the kinds of violations you consider major or serious, and another list of safety and health or other behaviors that, while not as serious, are still not acceptable. The following lists offer examples of both types.

Major offenses:

- Actions that demonstrate clear carelessness with company equipment or materials
- Horseplay in work areas or otherwise creating unsafe conditions
- Actions that demonstrate clear carelessness leading to an accident to oneself or to others
- Tampering with machine safeguards, removing machine tags or locks

- Not wearing required PPE
- Theft of company property
- Provoking or engaging in an act of violence against another person on company property
- Insubordination to proper authority
- Use of alcohol or illegal or nonprescription drugs that results in unsatisfactory performance
- Major traffic violations while using a company vehicle
- Other major violations of company rules or policies

General offenses:

- Wasting time
- Minor traffic violations while using company vehicles
- Creating unsafe or unsanitary conditions or poor housekeeping habits
- Threatening an act of violence against another person while on company property
- Misrepresentation of facts or falsification of company records
- Unauthorized use of company property
- Time and attendance problems
- Other violations of company policy and rules

You would then link each type of offense to a structured procedure for corrective action. Your goal here is to make sure that the corrective action is appropriate for the seriousness of the violation, that employees are given the opportunity to correct their own behavior, and that the system that directly links violations with corrective actions is used.

Written Warning:

- No safety glasses
- Horseplay

- Unsafe work habits
- Violation of other safety or health rule or regulation

Suspension:

- Three or more safety or health violations of the same type
- General overall record of unsafe practices
- Refusal to follow safety and health guidelines or instructions

Termination:

- Excessive and repeated safety and/or health violations
- Purposefully ignoring safety and/or health rules
- Unsafe actions that jeopardize the safety or health of others
- General disregard for safety and health of oneself and others

General Work Practices

These practices include housekeeping activities such as:

- Removal of tripping, blocking, and slipping hazards
- Removal of accumulated toxic dust on surfaces
- Wetting down surfaces to keep toxic dust out of the air

Safe Work Procedures

Each task of a job should be analyzed, the potential hazards in it identified, prevention or control of each hazard designed, and specific work procedures drawn up for it. Any manufacturer's operating instructions for equipment or Material Safety Data Sheets (MSDS) should also be used in developing safe work procedures. OSHA recommends involving workers who are familiar with the job in the analysis process. When properly followed, these procedures become the safe work practices that allow the worker to protect himself or herself from the hazard that remains after every feasible engineering control has been implemented. These usually involve a series of steps that must be taken to accomplish the job safely. They also include proper use and care of any PPE needed for the job. Where toxic dust is a hazard, they may also include personal hygiene routines.

Drawbacks to Work Practice Hazard Controls

While work practice controls are a necessity and can work very well, they are only as good as the management systems that reinforce them because they are vulnerable to human error. These hazard controls must be designed from the base of solid hazard analyses and must be accompanied by good worker training, reinforcement, and, when necessary, consistent and reasonable enforcement.

Work Practices Training

Training in safe work practices must begin with helping the worker understand why special precautions are needed. In effect, the training begins with a discussion of the hazard so that the worker understands the potential danger to himself or herself and to others. This can be difficult when new job procedures are being taught to workers who have done the job without special precautions long enough to feel no concern for the hazard(s). These workers must be helped to understand that a certain amount of luck is involved in avoiding injury or that illness may not show any symptoms for a long time after damage has begun.

Just presenting training may not be sufficient. An employer has a responsibility to ensure that worker training has achieved its objective, that the workers understand the hazards and know how to protect themselves. "Testing" the results of training can be done informally by the supervisor. This means stopping at a worker's work station and asking for an explanation of the hazards involved in the work and how the worker can protect himself or herself. If the training has been well presented and understood, each trained worker should be able to give a good, comprehensive response.

Positive Reinforcement

Each supervisor should reinforce the training in work procedures frequently. The informal "testing" described above serves as positive reinforcement as well as giving the supervisor needed information about the success of training. Some worksites also provide "recognition" for the use of safe work procedures. Sometimes this takes the form of little cards handed out periodically saying thank you for working safely and which can be redeemed for a free cup of coffee or soft drink. Another method is for the supervisor to watch individual workers at their tasks periodically and give them oral and/or written feedback on what was done safely.

Note: It is recommended that award systems be based on positive activities rather than on lack of injuries. Award programs with major "prizes" for hours worked without injury can put heavy pressure on workers *not* to report injuries.

Enforcement

It is important for workers to know that using safe work practices is a requirement of employment and that unsafe practices will not be tolerated. It is necessary, therefore, for the employer to have a disciplinary system that is fairly and consistently used if and when necessary. If no such system exists, the employer would be wise to have employees assist in the design and structure of one. Disciplinary systems are discussed in greater detail below.

Personal Protective Equipment (PPE)

A further way of controlling exposure to hazards when worker exposure cannot be completely engineered out of normal operations or maintenance work is to provide protective clothing or equipment collectively called personal protective equipment (PPE). This covers face shields (whether worn by dentists or welders), steel-toed shoes and boots, safety glasses and goggles, hard hats, leather aprons and belly guards, metal-mesh gloves, forearm guards, respirators, and "space suits." The full spectrum of what is available in PPE can be seen in the advertisements appearing in safety and industrial hygiene magazines or on display at the exhibitions at safety and industrial hygiene conferences.

Many OSHA standards require certain types of PPE. A section of the standards (29 CFR 1910, Subpart I) specifically addresses PPE. If respirators are *ever* worn for any reason at your worksite, OSHA standard requires a written respirator program.

If you are not sure what you are required to do or what might be best to do, you may call or write to the nearest state or federal OSHA office to ask for guidance. Employers in small businesses may contact the office of the OSHA-funded, state-run consultation service in their state.

The same drawbacks that exist for work procedures also exist for PPE in controlling hazards. Employees need good training in why the PPE is necessary and how to use and maintain it. They will also need positive reinforcement and fair, consistent enforcement. (See discussion below.) An additional drawback is that some PPE is uncomfortable and puts additional stresses on employees, making their work and their ability to work safely difficult. This is particularly true where heat stress is already a factor in the work environment.

By OSHA standard, employers are responsible for providing respirators that fit individual employees. Most employers provide required PPE, with the exception of safety shoes and safety glasses. Even in the latter two cases, employers usually pay for some of the cost.

Administrative Controls

Administrative controls are the controls of "last resort," when no other way is known to control hazards. They include rotation of workers through different jobs to "even out" exposure to hazards or to allow workers to work part of the day without respirators or other burdensome PPE. They also include longer rest breaks, additional relief workers, and the "exercise breaks" used to vary body motions during the workday.

Administrative controls such as these have been employed with dangerous but widely used toxic substances, including lead, with extreme temperatures and with ergonomic hazards. They are used in conjunction with other controls and should be replaced, when feasible, with better controls.

Hazard Correction Tracking

An important day-to-day part of hazard prevention and control is the correction of hazards that appear in spite of the overall hazard prevention and control program. Documenting the completion of hazard correction is equally important, particularly for larger sites. It is important for two reasons:

1. It keeps the Safety Department or the person in charge of safety and health aware of the status of completion for long-term correction items.
2. It provides a record of what occurred, should the hazard reappear or seem to reappear at a later date.

In addition, when an employee reports a hazard, the tracking of hazard correction allows correct and timely hazard correction information to be made available to the reporting employee.

Methods for Hazard Correction Tracking

Notations on the Report Form

Many companies use the form documenting the hazard "discovery" to track the completion of correction. That means that inspection reports would have notations about hazard correction alongside the information about the hazard. Employee reports of hazards and reports of accident/incident investigations would also have notations about hazard correction.

If your company is using forms or formats that do not specify a requirement to document the date of the completed action, you will, unfortunately, get documentation only of an *intended* action as often as of the *completed* action. Generally this will not pose a problem if the hazard can be corrected in a short period of time. Someone's memory will probably work to see that the final correction occurs. There is always a danger, however, that the expected correction will "slip through the cracks," usually when a part has to be ordered and time is needed to complete the procurement action.

Tracking by Committee

Some companies separate the tracking of hazard correction from the system that "discovered" the hazard. The most frequent form of separate hazard tracking is done through committees, which are either central management safety committees or joint employee-management committees. A few of the central committees have some employee representation. The mechanism for hazard tracking is frequently through review of inspection, employee hazard, and accident/incident reports. A listing of uncorrected hazards is made part of the committee meeting minutes. Thus part of each meeting is devoted to checking on completion of outstanding items.

The benefit of such a system is the high level of scrutiny applied to hazard correction tracking. The drawback is that since information has to be shifted from the reports to the committee agenda, there is a possibility that the uncorrected hazard will not be transferred over to the committee for scrutiny or that incorrect information will be provided to the committee. In most cases, however, since the committee is reviewing the original reports, information transfer should not be much of a problem.

Separate Tracking System Forms

Another way to track hazard correction is to transfer information from the reports to a separate hazard tracking system report. Ideally, this system receives *all* uncorrected hazards and not just the hazards from one of the systems of "discovering" hazards. Where possible, this system will be most effective if it is computerized.

For small businesses that do not use written reports for inspections or employee reports of hazards, this system allows documentation of hazard correction that otherwise might not exist.

The drawback is much the same as that for the committee tracking system in that there is always a possibility that incorrect information—or no information—will get transferred. (A form that can be used for this system is found in Appendix 7.)

Role of Disciplinary Systems in the Workplace

The primary purpose of a disciplinary system is not to punish your employees, but rather to control the work environment so that workers are protected and accidents do not happen. A disciplinary system helps ensure workplace safety and health by letting your employees know what is expected of them and providing them with opportunities to correct their behavior before an accident happens. In this way, a disciplinary system works along with other components of your safety and health program to maintain a high level of workplace protection and accident prevention.

A disciplinary system is one of the keys to implementing your safety and health program successfully. It ensures that your rules and safe working procedures are taken seriously by your employees and are put into use at the workplace. It lets your employees know how you expect them to operate in relation to the goals of your safety and health program, and what actions you will take if they do not meet your expectations.

A disciplinary system cannot work in a vacuum. Before you can hold your employees accountable for their actions, you first need to establish your safety and health policy and disciplinary rules. Then you need to develop safe operating procedures, train your employees on these procedures, and supervise their actions. (If you are starting a disciplinary system for the first time, the worksheet in Appendix 8 will help you plan what needs to be done.)

Policy Statement

Employees need to know clearly what is expected of them and what to expect if they do not do it. This is true in all areas of work but is especially important for worker safety and health. As part of the policy statement, or in an employee manual or booklet, you should have a written statement explaining your disciplinary policy.

Employee Information and Training

In addition to the written statement of your disciplinary policy, there should be a list of what you consider major violations and less serious violations of company policy, and the kinds of disciplinary actions that will be taken for first, second, or repeated incidents.

It is important that employees know and understand the system and have a ready reference to turn to if they have any questions. Training that will reduce the need for disciplinary action includes instruction on the importance of workplace safety and health, the need to develop safety habits, safe operating procedures and the hazards they control, and the standards of behavior that you expect. You will want to ensure that your employees understand your disciplinary system and the consequences of any deliberate, unacceptable behavior.

Supervision

Supervision incorporates both training and corrective action. Ongoing monitoring of your employees' work and safety habits gives you and/or your supervisors an opportunity to observe how work is performed and to correct any problems before a serious situation develops. Effective supervision of this nature allows correction before any punishment, and an open, interactive relationship between employees and their supervisors that allows them to discuss problems and reach mutually agreed-upon solutions. This type of relationship creates a work environment where the need for disciplinary action is reduced, and if it is needed, it is perceived as corrective rather than punitive.

Employment Involvement

You may want to involve your employees in setting up or revising your disciplinary system. When employees have an opportunity to contribute their ideas to the establishment of workplace rules and to setting appropriate disciplinary actions, they will be more knowledgeable about the system and more aware of how it is designed to protect them against the unsafe acts of others. Of course, at sites with collective bargaining agents, you will need to involve employee representatives.

Employees should be encouraged to help informally in the enforcement of rules and procedures. This can contribute greatly to the success of a disciplinary program. This is not to say that you would want your employees spying and reporting on each other, but rather that you can encourage them to be one another's "keeper," watching out for the safety and health protection of their fellow workers. Many employers have successfully encouraged the creation of a company "culture" where employees are comfortable telling someone when they spot a problem that can easily be corrected, such as reminding a co-worker to put on safety goggles.

One other important area of employee involvement is the opportunity for employees to correct their own behavior problems. An effective disciplinary system is a two-way process. Once a problem is spotted, it should be discussed with the employee and he or she should be given at least one or two opportunities to change the behavior or correct the problem. Only after these discussions (and, perhaps, some retraining) would you want to take disciplinary action.

Appropriate Control Measures

Any disciplinary action taken needs to be appropriate to the importance or hazardousness of the problem and the frequency with which it occurs. It is certainly inappropriate to fire someone for occasional tardiness, but it is also inappropriate to issue only oral warnings to an employee who repeatedly removes a machine guard. There is a correlation between rules and the sort of discipline that might result from breaking them.

Consistent Enforcement

If your disciplinary system is to work well and is to be accepted by your employees, every employee needs the assurance that it applies equally to all. You will also want to make it clear that managers and supervisors are subject to similar rules and regulations, and subject to similar or even more stringent disciplinary actions.

It is also important in the administration of corrective procedures to provide feedback to the employee on why the corrective action is necessary, why the employee's action or behavior is unacceptable, and how the employee can prevent future violations and discipline. You will also want to recognize employees for improving or correcting their own behavior.

Documentation

One key to ensuring fairness and consistency in a disciplinary system is to keep records of corrective actions taken. It is in the best interest of both the employer and the employee to have written rules and disciplinary procedures, to keep records of problems when they are noted and discussed, and to maintain personnel files that include documentation of good or poor safety and health behavior. In this way, employees are kept apprised of any problems that need correction, and employers are able to track corrective actions and their impact. The latter is also important when employers are evaluating the management and supervision skills of their supervisors.

If a safety and health behavior problem is so bad that warnings, retraining, and other corrective actions have had no effect, then documentation is critical before discharging an employee. You may want to consider clearing the files annually when employees have overall safe work records and the warnings received are minor in nature.

Reward System

If it is important to correct or even to punish unsafe behavior, it is at least as important to reward safe behavior. A reward system provides a good counterbalance to a disciplinary system by focusing on the positive and thereby motivating employees to practice safe and healthful work habits. It is especially important to recognize self-motivated acts of safety and/or health protection—when employees take the initiative to protect themselves or others.

A reward system can be very simple and inexpensive: letters or certificates of appreciation, a few hours of paid leave, a special and convenient parking space for a month in the company parking lot, a small pin or tie tack. Reward programs should be evaluated periodically and refreshed with new ideas if employees show signs of losing interest.

Any public recognition and rewarding of good safety and/or health behavior will both recognize the employee and provide incentive to other workers. The recognition is more important than monetary value when assigning one-time awards. Of course, taking safety and health performance into account when promoting or providing bonuses to employees is probably the most meaningful reward for good safety and health performance.

Note: Award/reward programs based on lack of accidents can do more harm than good if they put pressure on employees to underreport injuries and illnesses. Beneficial award/reward programs should be based on constructive efforts of employees.

A company disciplinary system is rather like a life jacket: you don't want to have to use it, but you want it to be there in case you need it. In addition, a disciplinary system can have a positive impact. It lets your employees know just how concerned you are about safety and health. It tells them what behavior you expect of them. It assures them that corrective actions will be fair and the same for all. It lets them know that you are trying to protect them against the unsafe acts of others. For you and your safety and health program, a disciplinary system is a key link between the policy and rules you developed and the results you want to achieve in production, in profits, and in workplace safety and health.

Preventive Maintenance

Although not frequently thought of as part of a safety and health program, good preventive maintenance plays a major role in ensuring that hazard controls stay in place and remain effective as well as prevent new hazards from arising due to equipment malfunction.

Whenever systems are enclosed, the enclosure usually depends upon the smooth functioning of pipes, valves, pressure releases, and so on. Malfunctioning of these parts of the enclosed system may result in hazards to workers.

Where ventilation systems provide hazard controls, stability of these systems to provide that control relies upon the smooth functioning of duct work, fans, and filters. Many guards are electronic or electrically energized and require maintenance to continue to operate smoothly. In addition, equipment that may not be hazardous under normal conditions may become so when it begins to malfunction. Preventive maintenance keeps that from happening.

Preventive Maintenance Scheduling

Preventive maintenance requires good scheduling of maintenance activity. The schedule is in turn dependent upon knowledge about what needs maintenance and how often. The whole point of preventive maintenance is to get the work done before repairs or replacement *must* be done.

Maintenance Needs Survey

A preventive maintenance program starts with a survey of maintenance needs at the worksite. Every piece of equipment or part of a system that needs maintenance—such as oiling, cleaning, testing, replacement of worn parts, or checking—should be surveyed, and all the parts to be maintained should be listed. If you do not have such a complete and thorough listing at your worksite, the supervisor for maintenance should be required to develop one. The survey should be redone periodically and the list of items updated. Whenever new equipment is placed in the worksite, the list should be revised to take that into account.

Maintenance Timetable Determination

Once the complete list is developed, a timetable must be established. A determination should be made as to the average length of time before the maintenance work becomes reactive rather than preventive. Then the maintenance action should be planned for a time in advance of that average time. (Maintenance should be performed at least as often as recommended by the manufacturer.) Maintenance documents should be reviewed periodically to see how much reactive maintenance (repair or replacement of defective parts after failure) has been done. New estimates of average time should be made and the schedule adjusted accordingly.

Posted or Computerized Schedules

The schedules for maintenance should be easily and generally accessible at the worksite. This helps keep the maintenance employees knowledgeable in planning their work and helps hold the maintenance department accountable for following that schedule.

Maintenance Documentation

Because preventive maintenance can be a complicated matrix of timing and activity, it is easy to lose track of in most workplaces without some form of documentation of completed activity. In most workplaces, this documentation is a simple addition of date and initials to show completion on the original schedule of work to be done. Other workplaces have information about who did the work, the date it was done, and any other information entered into the computer system. Documentation also assists worksite management in holding those responsible accountable for the successful prevention of costly repairs or accidents/incidents.

Maintenance of equipment and systems is necessary to ensure the continuation of hazard controls and the prevention of new hazards caused by malfunctioning equipment. Preventive maintenance is achieved by implementing a schedule of maintenance that forestalls malfunctions. Prominent posting or computerization (with general viewing access) makes clear what needs to be done, when, and sometimes by whom. Documentation of maintenance work allows working accountability systems to be established.

Emergency Preparation

The Nature of Emergencies

Emergencies are situations in which hazards not normally in the workplace appear. The emergency-spawned hazards may be the result of *natural disasters* such as earthquakes, tornadoes, hurricanes, floods, or ice storms. They may also occur due to *man-made* events outside of your control of the worksite, such as train or plane accidents, terrorist activities or events at nearby worksites that impact on your worksite. Finally, emergencies may occur within your own systems due to *unforseen combinations of events* or the failure of one or more hazard control systems.

Emergencies, by their nature, are not expected as part of the routine, as if they happened every day. They may never occur, but if they do, their cost both in terms of dollar losses and human suffering can be enormous. Your job is to understand what emergencies are possible, rather than just probable, and to plan the best way to control or prevent the hazards that they present.

Surveying Possible Emergencies

Your emergency preparation should begin with a survey of all the possible emergency situations that could occur at your worksite. Just because a certain type of emergency has never occurred does not mean that it never will in the future. Start by listing all the possible emergency occurrences in each of the three categories, taking into consideration the unique characteristics of your worksite and its location.

Natural disaster

Review each type of natural disaster that has been experienced in your geographical location and check with experts on the chances of other types of natural disasters occurring in your geographical area.

Man-made Events or Deliberately Caused Disasters Beyond the Controls of Your Worksite

Consider the environment of your worksite. Are you near an airport or on the takeoff and landing pattern for an airport? Is there a train track carrying products (other than those that you ship or receive) near enough so that an accident to a train carrying toxic materials might impact your worksite? Are there chemical or other dangerous worksites in your neighborhood that could have internal emergencies might could impact your worksite? Have there been terrorist activities against plants belonging to your company or involved in similar processes or producing similar products?

Hazard Control Failures at Your Worksite

Ask yourself what are the worst things that could possibly happen as a result of conditions here? Every worksite has some potential for fire; some have much greater potential than others. What is your potential for explosion or release of toxic substances?

Emergency Planning

Once you have listed all the possible emergencies, you should plan actions to reduce the adverse impact of the emergency on the safety and health of your workers. Some actions will be the same or similar for all types of emergencies. Some may require quite different or even opposite actions. You will need to plan what kind of first aid or medical response is needed and where that response will come from. If you are relying on outside medical or emergency response organizations, you will need to establish communications with them and plan together for emergencies. If possible in those situations where you are working with outside resources, it is advisable to have those resources participate in your drills.

Employee Information and Training

After planning has been done, your employees need to be informed of those plans to the extent that they may need to act on them. Each employee needs to know precisely what he or she is expected to do in each type of emergency as soon as they are alerted.

For the most probable emergencies, employees should be drilled in the actions you expect them to take so that it becomes second nature to them. In this way, if and when an emergency occurs, they will be able to do what they must to protect themselves and others in spite of the heavy stress that the emergency is sure to bring about. Fire and evacuation drills should be held annually. Other types of drills, such as tornado or earthquake drills, should follow a predetermined schedule based on frequency and probability of need.

Emergency preparation and planning should begin with a list of possible emergencies based on the location and environment of your worksite. Where outside organizations will be needed for your emergency response, planning should be done jointly. Your employees should be informed, trained, and drilled in various aspects of emergency preparation.

Medical Programs

The term *medical programs* can be confusing. It is intended here to refer to the way occupational health care is provided. Medical programs consist of both occupational health approaches to identifying health problems that may be work-related, and responses to injuries and illnesses that occur. The size and complexity of a medical program will depend on the size of the worksite, its location in relation to health care organizations, and the nature of the hazards at the worksite. (See Chapter 11 for more information about medical programs.)

Medical programs consist of everything from first aid and CPR response to sophisticated approaches for the diagnosis and resolution of ergonomic problems. The extent of your medical program and the way you provide for that program will depend on a number of factors. You must always be prepared to offer first aid at your worksite. In fact, this is required by OSHA standard (29 CFR 1910.151 [b]) for worksites that are not close to medical facilities. OSHA strongly advises that both first aid and CPR assistance be available on every shift at your worksite.

Employers in small businesses may contact the OSHA-funded, state-run consultation service in their state for assistance in deciding what type of medical program is needed at their sites. If use of nearby medical facilities appears to be the best arrangement for your worksite, be sure to meet with representatives of that facility to discuss your worksite's medical needs.

Please note that whatever the configuration of the medical program you use, it is important to use medical specialists with occupational health medical training. Not every nurse or doctor is trained to understand the relationships among the workplace, work, and certain medical symptoms.

SUMMARY

Each category of hazard in the workplace should be approached with the concept that it will be totally prevented, if feasible. If total prevention is not feasible, the hazard should be controlled as completely as is feasible through work and equipment design. To the extent that the designed controls allow potential exposure, safety and health rules, work procedures, and PPE should be used to control that exposure. If, after all feasible use of these methods has been exhausted, some hazard exposure still exits, administrative controls should be used to reduce that exposure for individual workers.

To complement these hazard controls, good systems of preventive maintenance, hazard correction tracking, and fair, consistent enforcement of rules, work procedures, and PPE must also be used. Emergency preparation provides for the control of hazards that are unexpected as part of routine operations, and the medical program helps with both the identification of hazards and the minimization of harm when it occurs. All of these are part of a system of hazard control and prevention that provides comprehensive worker protection from occupational hazards.

11 The Occupational Health Delivery System: A Medical Program That Is Right for Your Worksite

The goal of the Occupational Safety and Health Act of 1970 is "to assure so far as possible every working man and woman in the Nation safe and healthful working conditions" (29 U.S.C. 651, Sec. 2 [b]). The identification and control of health hazards and the implementation of a medical program is extremely important to a total safety and health program.

A medical program is another name for the systems that employers put in place to ensure that occupational health expertise is used in their overall safety and health programs. Having a medical program does not necessarily mean that you must go out and hire a doctor to work at your company. There are many ways for you to find and use occupational health expertise. This chapter is aimed at helping *you* decide what will work best for *your* business.

This chapter refers to the medical program as the "occupational health delivery system," or OHDS. This term will help you remember that a comprehensive program is more than a *response* to work-related injuries and illnesses *after* they happen. It also means those activities that you do to find the safety and health hazards in your business and your plan for their prevention or control. It is a management system in the same way that the activities you do to promote safety and health are a management system. After all, your safety and health system is much more than the office occupied by the person in your company who is responsible for safety and health. Similarly, your OHDS is more than the place in your business where you keep the first aid kit.

If you find it more difficult to establish the goals and objectives for your OHDS than for the other parts of your safety and health program, it may be because the harm it prevents may not seem obvious at first. For example, an employee who has hand pain and is gradually developing a cumulative trauma disorder (CTD) will seem to have a less serious problem than the employee who has a severe cut or a broken bone from an accident. But OSHA's experience is that work-related health problems are no less serious in terms of costs and human suffering than the more obvious injuries.

An effective OHDS will help reduce all types of safety and health hazards and the resulting injuries and illnesses. The positive results from such a program will be measurable by a decrease in lost work days and workers' compensation costs. They will also help increase worker productivity and morale.

Who Should Manage the OHDS?

You will find that having occupational health professionals (OHPs) manage the OHDS works best. This means a physician or a registered nurse with specialized training, experience, and knowledge in occupational health works with you but *not necessarily as your employee.* Such an arrangement works best because safety professionals, industrial hygienists, occupational medicine physicians, and occupational health nurses each have their own area of specialized knowledge. You cannot expect to get all the information and service you need for your safety and health program from only one type of specialist. If you tried, you might miss identifying a dangerous hazard in your business.

This chapter describe the different ways that physicians and registered nurses receive specialty training in occupational medicine and health and the different services that they can provide for your business.

What Services Do You Need from Your OHDS?

What to Consider

There is no such thing as a standard OHDS. As an employer or manager for a business, you need to think of the special characteristics of your business when developing an OHDS that is right for you. These special characteristics include:

- The actual processes in which your employees are engaged
- The type of materials handled by your employees
- The type of facilities where your employees are working
- The number of employees at each site under consideration
- The characteristics of your work force, such as age, gender, ethnic group, and education level

- The location of each operation and its distance from health care facilities

As you look at the characteristics of your worksite, you should be asking yourself questions such as:

1. Are there hazards in our processes, materials, or facilities that make it likely that our employees will get sick or hurt?
2. Do we have so few employees that on-site occupational health resources are less practical than off-site contract services? Do we have so many employees that we will save time and money by installing on-site resources?
3. Are there special characteristics of our workers that would make them more vulnerable to illness or less likely to understand the dangers of illness?
4. Is there anything about our workplace that makes it important to have occupational health assistance nearer to us or faster to reach?

When you examine your business with these questions in mind, you will be better able to choose which OHDS services you need. The services are listed below.

The Full Range of OHDS Functions

There are three basic types of OHDS activities:

1. Prevention of hazards that cause illnesses and injuries
2. Early recognition and treatment of work-related illness and injury
3. Limiting the severity of work-related illness and injury

Prevention of Hazards

1. Make sure that your safety and health policy shows that you are as concerned about your employees' health as about their safety.
2. Make sure that qualified OHPs help you identify the hazards and potential hazards of your workplace.
3. Use OHPs in the development and presentation of health training.
4. Provide professional occupational health expertise as a resource to your safety and health committee.
5. Be sure to include your OHDS in your annual self-evaluation to ensure the quality it provides.

Early Recognition and Treatment

1. Use OHPs to help you decide, on the basis of existing or potential hazards at your workplace, when base-line and periodic testing of your employees and new hires for exposure evidence may be necessary. This is called "health surveillance" and is required by some OSHA standards for specific types of exposures.
2. Use OHPs to do the testing needed for health surveillance.
3. Make sure records are kept of employee visits to first aid stations, nurse's office, contract clinic, or hospital and have an OHP review the symptoms reported and the diagnoses to see if there appear to be patterns that indicate an occupational health problem.
4. Provide first aid and CPR assistance through prop-

erly trained employees on every shift, making sure that employees keep up their certifications and that they are adequately trained in the specific hazards of the worksite. Also make sure that they understand the hazards from blood-bearing communicable diseases, including hepatitis B and AIDS, and how to protect themselves.

5. Make sure that the OHPs whom you use have current credentials, have had recent occupational health continuing education, and understand the hazards of your worksite, so that they will be able to recognize *early* symptoms of occupational health problems and begin prompt and appropriate treatment to prevent disability.

6. Make sure that standardized procedures (called "protocols" in the medical community) are used in all aspects of occupational health delivery systems, particularly if you are using more than one contractor for health services.

7. Have one of the OHPs keep your employee injury and illness records.

Limiting Severity

1. Coordinate the response to emergencies of all the responsible individuals or departments that are at your worksite and of all the emergency organizations off the worksite, such as the fire department, any contractual organization, or a nearby community hospital, so that everyone knows exactly what to do and what to expect from others.

2. Maintain contact, through your OHP (whether your employee or arranged by contract), with any employ-

ee who is off work due to an occupational illness or injury, keeping in touch with the practitioner providing treatment and care to ensure that the treatment is appropriate and that the employee is responding as expected.

3. Use your registered nurse or physician contact with an employee who is off work for an extended period to help advise the employee about workers' compensation benefits and rights and about ongoing care.
4. Use these OHPs to provide evaluations aimed at determining whether an employee can resume full duty after injury or illness or whether their work duties need to be modified.
5. Consult your physician or registered nurse for help with the development of the modified duty position, to ensure that the employee can perform the work and get the benefits of feeling productive again.

Examples of OHDS Use

Following are three examples of the organization of the OHDS at large, medium, and small workplaces. The first is a manufacturing firm with about 1800 employees, divided between two locations. The second is a small, independent janitorial service with 35 employees. The last example is a meat packing company with about 500 employees. These examples show that each company's service needs are individual and that their methods for providing the services differ. The key issue demonstrated in the examples is that sometimes services are best provided by people that you hire as your own employees; in other situations it may be better to contract for these services. You will notice that sometimes employers in our examples used community agencies to supply services; this may save you money.

Background Information on the Sample Companies

Example 1

One thousand of the workers employed by the large manufacturing company are at one site, where the components of their major product are manufactured. A second site, five miles away, has 500 employees and includes a line where the components are assembled. Also in that building is the facility's maintenance department and a garage where the company trucks and other vehicles are parked and serviced. At that same site but in a different building are the executive offices, which house all the administrative divisions, such as accounting, human resources, marketing, and a small showroom.

The manufacturing site is seventeen miles from the nearest medical facility. The light assembly and administrative site is twelve miles from the nearest medical facility. The company runs two eight-hour shifts: a day shift and an evening shift. There are two security personnel working alone from midnight to 7:00 A.M.

Company one employs a safety professional to head its safety department. This company employs an occupational health nurse-practitioner to head its OHDS. Both departments have other professional and nonprofessional staff to support the department heads. The safety department and OHDS both have offices at the manufacturing site.

Example 2

Company two, the small, independent janitorial agency, provides cleaning and light maintenance services for commercial buildings. There are 35 employees, including the owner-manager. There are three clerical support personnel, an evening supervisor, and 30 service personnel, for a total of 20 men and 10 women. The service personnel report to a central office from which they are dispatched in company vans in teams of two or three. The service personnel all work an evening shift from 5:00 P.M. to 1:30 A.M. Two of the clerical employees work a day shift, 9:00 A.M. to 5:30 P.M. One clerical employee works a shift that spans the day and evening shift. The supervisor works the same evening shift as the service personnel.

No safety, industrial hygiene, or occupational health professionals are employed by this company.

Example 3

Company three, the meat packing company employing 500 workers, has 460 people working in the slaughter department and the department where the carcasses are dismantled. These divisions operate on a ten-hour per day shift. In addition, there are 20 people who perform plant sanitation and maintenance functions on an overlapping evening shift. There are also 20 employees functioning as supervisors and in staff positions such as personnel, payroll, and safety and health.

The plant is seven miles from the nearest health care facility. The company employs a full-time occupational health nurse (registered nurse) and a full-time safety director.

In the sections below you will find major portions of an OHDS and descriptions of the way they are handled by each of our three sample workplaces.

Hazard Analysis

At *company one*, the basic work to provide a comprehensive assessment of hazards was done by a committee composed of the safety director, the occupational health nurse practitioner, the director of manufacturing, the supervisor of maintenance, and two line employees, one from the day shift and one from the evening shift. In addition, a consultation was requested from the liability insurance company's loss control division. The consultation was conducted by an industrial hygienist who confirmed the need to monitor for noise in the manufacturing area. She also helped the occupational health nurse practitioner and the human resources director write job descriptions for the major employee classifications. The descriptions emphasized important safety and health considerations such as the amount of weight lifted and the chemicals handled.

A second consultation was obtained from the department of occupational medicine at the nearby university. The occupational medicine physician suggested a design for an optional health surveillance program for cumulative trauma disorder (CTD) in the shoulders, arms, and hands of the employees in light assembly. The program is designed to use personnel and other resources already available at the company.

At *company two*, the owner-manager of the janitorial service was aware, from reading the newspaper, that OSHA was enforcing protective measures for workers who were exposed to ergonomic hazards. She contacted the OSHA-funded, state-run consultation service and received information about the criteria that were being used for enforcement. This information guided her in organizing a management system that would meet the enforcement requirements. At the owner's request, consultation personnel examined this system and other aspects of the company's safety and health program, including the hazards to which employees were exposed. They helped develop protections against the hazards found, and recommended a cumulative trauma disorder (CTD) training program with an emphasis on back injuries. They also recommended a driver safety program.

At *company three,* the meat packing company's hazard assessment included basic safety and industrial hygiene evaluations as well as a complete ergonomic review following the *OSHA Ergonomic Program Management Guidelines for Meat Packing Plants.* Their assessment included a check for potential exposures to communicable diseases from the slaughtered animals. The safety and health committee developed a list of questions about safety and health conditions and potential hazards in the plant. To answer these questions the committee performed a series of walk-throughs and employee interviews. The committee consisted of the safety director, the occupational health nurse, a supervisor, and four hourly employees, one each from the slaughter, fabrication, sanitation, and maintenance departments.

In addition, the safety director and the occupational health nurse analyzed the actual jobs being performed by line employees for potential ergonomic problems. The OSHA ergonomic guidelines helped them identify those positions involving the specific activities associated with the development of CTD. Furthermore, they reviewed all of the material safety data sheets (MSDS) for all the chemicals used for cooling and sanitation at the plant.

Employee Training

At *company one,* the safety director and the occupational health nurse practitioner designed training programs to be delivered during employee orientation to inform the employees about the company safety and health policies and program, and to alert them to the specific hazards in their jobs and what they need to do to protect themselves. The training is conducted by safety department and

OHDS staff members who have taken special courses to prepare them as trainers. Employees are invited to contact the safety director, the occupational health nurse practitioner, or their supervisors if they have further questions. The training includes a short test at the end to demonstrate that the employees understand their specific risks and how to protect themselves. There is a regular schedule for follow-up training.

Because the plants of company one are more than ten minutes away from the nearest medical facility, the decision was made to establish emergency response teams in both locations and on both shifts. The teams, set up on a volunteer basis, consist of five employees per shift per location and all the security personnel. The company contracted with the American Red Cross to provide the first aid and cardiopulmonary resuscitation (CPR) training and refresher classes at the company.

At *company two*, the state OSHA consultation office suggested contracting with a nearby occupational health clinic, already providing work-related employee health services, to conduct training in the prevention of CTD with an emphasis on back safety. In addition, the employees are to be informed about the contents of the cleaning solutions they are using, along with proper mixing techniques and the use of gloves and protective eye wear. The MSDSs, which list toxic ingredients, are explained and employees are told where the MSDSs are kept. The initial class was provided at the janitorial service company's central office. New employees will be instructed at the clinic as part of their preplacement physical examination and orientation. Finally, the owner received booklets from the Automotive Occupant Restraint Council promoting safe defensive driving and the use of seat belts. These booklets were distributed to all current employees and are included in the orientation materials for all new employees.

By granting paid time for the class, the employer encourages service personnel to take a beginning first aid course offered through a local municipal adult education program. The evening supervisor was required to take the beginning and advanced first aid course. Each van is supplied with a first aid kit, as is the office. Service employees are instructed to report all injuries and illnesses to the supervisor at the end of their shift, and sooner by phone if they think the problem needs more than minor first aid treatment. The supervisor then refers the employee to a nearby emergency room or the contract occupational health clinic as necessary.

Management in *company three* decided to conduct the employee training program using their own company personnel. The safety director designed the training to address hazards such as fire, walking surfaces, cuts, elevations, and ammonia leaks. The occupational health nurse designed training to address hygienic practices to reduce the possibility of exposure to such biologic hazards as brucellosis, anthrax, and Q fever associated with animal handling. In many respects, the activities performed by the workers to protect the meat from contamination, as required by the USDA, also protects them. Where this is not the case, the program is designed to emphasize what employees need to know and do to protect themselves. The nurse also developed material to inform employees about the early signs and symptoms of CTD and ways to help prevent them. Together, the occupational health nurse and the safety director at the meat packing company instruct the employees about the chemical and temperature hazards associated with the industry. Additional classes are held for the clerical, sanitation, and maintenance workers. In addition, supervisors are trained about the early symptoms of ergonomic problems so they will encourage workers to report these problems as readily as other injuries or illnesses.

Five volunteers from each plant area or unit and all the

supervisors from each shift were trained in first aid. Commitment to be on the emergency response team is for one year. Retraining will occur for all team members at the anniversary date, when new members are added.

Health Surveillance

In *company one,* machines in two departments were extremely noisy. The safety director designed and the maintenance department constructed double-layered sheetrock walls with sound-reducing baffles between them around the two machines. The company then contracted with an industrial hygienist to perform an environmental sound survey, which showed that the noise level in one of the departments was still too high even after the construction of the sound baffles. The company also contracted with a nearby audiologist to perform baseline pure tone hearing tests on all current employees. New employees for the department designated as too noisy will be tested as part of their orientation. The required annual audiometric testing is done by this same audiologist. The occupational health nurse practitioner conducts the education program about hearing conservation. She also did the research necessary in order to purchase the best hearing protectors for the employees.

Company two had no need of health surveillance.

Company three consulted with the occupational medicine physician to develop a surveillance program for cumulative trauma disorder (CTD). As a result the physician made the portion of the preplacement physical examination that dealt with the upper extremities and the back more detailed for the workers in the slaughter department and the division where the carcasses are dismantled. At the six-month and one-year anniversary dates, randomly selected employees are invited back to be reexamined.

Examination results that indicate early development of CTD are reported to the employees and management is informed about which positions need further evaluation. However, no personal information that identifies particular employees is given to management.

Analysis of Employee Use of the OHDS

At *company one,* a clerk in the OHDS is assigned to enter information about each employee visit to the OHDS onto a spreadsheet that includes the date, the time of day, the employee's department, the employee's complaint, and the treatment rendered to the employee. Totals from the graph are examined by the occupational health nurse practitioner every month. Any unusual clusters of complaints are investigated by the safety committee.

At *company two,* the contract occupational health clinic sends a monthly statement to the employer summarizing all bills that have been submitted to this company's workers' compensation insurance carrier. This summary includes both diagnostic and treatment information. The employer analyzes this information for trends in injuries and illnesses as one way of deciding if employees are being exposed to identified hazards, if hazards exist that have not been identified, or if employees need more training about hazards.

At *company three,* the occupational health nurse maintains a spreadsheet of all employee visits to the health office. She combines this information with that received from the contract occupational medicine physician to form a report that is presented to the safety and health committee each month. This report and the accident reports become the basis for special safety and health emphasis programs within the company.

Establishing Standard Procedures

For *company one*, the occupational health nurse wrote procedures for the "first aiders" to use when administering first aid, CPR, and emergency transfer of ill or injured workers. She also wrote procedures describing the standardized assessment and on-site treatment that she uses for employee illnesses or injuries and for all health surveillance programs.

For *company two*, the state OSHA consultation staff worked with the owner to develop standardized procedures for first aid and emergency situations. They also wrote specific procedures for mixing and use of all the cleaning solutions, using the buddy system for lifting heavy objects, and a system for task rotation for lengthy repetitive activities such a vacuuming. Discussion of the procedures is included in the employees' orientation.

At *company three*, the occupational health nurse and the contract occupational medicine physician worked together to write procedures for all the treatment given in the OHDS office, including the dispensing of over-the-counter medications. Procedures were also written for hygienic practices for the employees exposed to biologic hazards that ensured that they did not consume food, beverages, or tobacco products with contaminated hands, and that they did not accidently contaminate their street clothes or shoes before leaving the plant. The safety director and the occupational health nurse developed procedures to ensure that the employees of contractors performing pest eradication operations did not accidently expose employees to pesticides. Finally, this same team developed procedures that included proper work techniques and frequent knife sharpening to prevent ergonomic problems.

Early Recognition and Treatment

At *company one,* the occupational health nurse-practitioner who heads this department has graduate level training in assessment and management of occupational illnesses and injuries. She is licensed to treat many of the employees' occupational injuries and illnesses, using previously approved standard procedures. A referral relationship has been established by contract with a local hospital that has an emergency room and an occupational medicine clinic. Employees with illnesses or injuries that are assessed by the nurse practitioner to be too severe to be treated on-site are transferred to the emergency room. Those employees who are receiving treatment from the nurse practitioner and do not respond as expected are referred to the occupational medicine clinic. In this way, a majority of the company's work-related injuries and illnesses are treated within the OHDS.

At *company two,* the contract clinic's medical director has completed a mini-residency in occupational medicine and has ample knowledge of the risks to which this company's workers could be exposed.

Company three has a policy that encourages employees to promptly report symptoms of illness and injuries to the occupational health nurse. The nurse treats minor illnesses and injuries in the plant health office using simple dressings, ice, and over-the-counter medications, and refers more severe problems to the contract physician. This nurse has taken a continuing education course in the recognition and conservative treatment of CTDs and is able to implement treatment and referral early. She is also able to review preventative measures with the employees at each visit.

Medical Case Management

At company one, the occupational health nurse practitioner works with the personnel manager to develop a case management system for all employees who are off work with illnesses or injuries lasting more than five days. The system consists of a method for prompt treatment authorization, a referral list for second opinions, assistance in filling out insurance forms, communication with the insurance carrier to ensure timely benefit payments, and ongoing contact with the employee and the family to answer questions.

At *company two,* the owner functions as the human resources director as well as the manager. As such, she is in close contact with any employee who experienced lost work time related to industrial injury or illness. She considers this adequate case management. Her workers' compensation insurance carrier assisted her by providing information about helping injured and ill workers return to work quickly. The suggestions prompted her to increase the frequency with which she makes telephone contact with those persons.

At *company three,* the occupational health nurse practitioner and the contract occupational medicine physician maintain close communication about all employees with work-related injuries and illnesses that are not responding to treatment as expected. The occupational health nurse practitioner ensures that specialist referrals occur promptly. The practitioner also works closely with the supervisors so that employees can return to modified duty positions as soon as possible by clearing the proposed positions with the treating physician.

Coordination for Emergency Services

At *company one,* the safety director, the occupational health nurse-practitioner and the head of the security department worked together to develop a system whereby all employees in each department know their exact responsibilities in the event of an emergency. They also discussed their plan with the local fire department that will be responding to emergency calls.

At *company two*, because of the nature of the operation, with employees moving from one workplace to another instead of having a fixed worksite, no special arrangements were made with emergency organizations. The company, however, did make its own emergency preparations (discussed under "Training.")

At *company three*, the occupational health nurse-practitioner and the safety director head the team and respond to each emergency. The plant receptionist is responsible for contacting outside emergency organizations, so he or she is included in the emergency response team meetings.

Recordkeeping

At *company one*, the OSHA 200 Log, the MSDSs, and the results from the noise surveillance are maintained and kept in the OHDS office, where the occupational health nurse practitioner can answer employee questions. All employee visits to the OHDS are documented in the individual employee medical record.

At *company two*, the OSHA 200 Log and the material safety data sheets are maintained by the owner and are available for the employees to see upon request.

At *company three*, the OSHA 200 Log, the MSDSs, and the results from the CTD surveillance are kept in the OHDS office, where the occupational health nurse can answer employee questions. All employee visits to the OHDS are documented in the individual employee medical record.

The examples demonstrate how three different employers provide OHDS services using a combination of resources within their companies, subcontractors, and governmental agencies. Some of the subcontractors brought the services to the premises, while in other situations the employees traveled to the contractor to receive the service. In each case, the services provided were selected based on the special characteristics of the business process, the potential exposures within that process, the business location, and the employee population.

The OHDS in the examples also demonstrate the use of activities aimed at the prevention of exposures, the early recognition and treatment of work-related illnesses and injuries, and those activities that decrease the severity of and potential for disability from work-related illness and injury. The inclusion of all three types of activities produce the safest and most healthful work environment.

Recordkeeping

OSHA requires that you keep certain records. Employee visits to the health office will generate other records. Maintaining your records using the resources of your OHDS is a good option because the occupational health professionals connected with your OHDS have the training to answer your employee's questions about these records and decide whether their complaints are work-related. Please remember, in addition, that individual employee health records contain confidential information that may be protected under case law relating to the right of privacy. Access to these records should be managed by health care professionals.

What follows is a description of the health-related records that you should have in your business. It is meant to give you enough information to decide if you want to manage these records within your OHDS. You need to consult the standards or the other references for more details. The Code of Federal Regulations (CFR) numbers or publication titles have been included for your convenience.

OSHA-200 Log

OSHA requires that you keep track of all the work-related injuries and illnesses that occur at your work sites. OSHA Form 200, often called the OSHA-200 Log, is used for that purpose. Another form, the Supplementary Record of Occupational Injuries and Illnesses (OSHA form number 101), or its equivalent, must also be kept. Employers often use the "First Report of Injury," required by the workers' compensation system, as this supplementary record.

The OSHA Recordkeeping Standard (29 CFR 1904) also tells you how long you must keep the records, and how you are to inform your employees about the injuries and illnesses occurring in your company. For more information, see this standard and the Bureau of Labor Statistics 1986 guidelines, "A Brief Guide to Recordkeeping Requirements for Occupational Injuries and Illness" or the more comprehensive "Recordkeeping Guidelines for Occupational Injuries and Illnesses" (both are OMB 1220-0029).

Environmental and Employee Health Monitoring Records

When you do any type of environmental monitoring, such as air sampling, OSHA requires that you keep the results. If you test employees for exposure to hazards, you must keep this information also. The publication "Access to Employee Exposure and Medical Records" (29 CFR 1910.20) tells you that employees and their representatives must be allowed to see and copy the records, and tells you how long the records must be kept.

Records Required for Specific Hazards

OSHA has issued a number of standards explaining what you must do when your business involves certain hazards. Often this includes testing the workplace and the employees for signs of the hazard. When this is the case, the standard also tells you how long you have to keep the results of the tests. Two examples of this type of standard are the asbestos standard (29 CFR 1910.1001[m], and the occupational noise exposure standard [29 CFR 1910.95].

Hazard Communication

The Hazard Communication Standard addresses more than recordkeeping. It tells you how you must communicate to your employees and your community about the chemicals you use or make in your business. But because the standard requires that you keep certain documents and communicate certain information, it is discussed here.

The standard says that you must, among other things, keep a list of all hazardous chemicals present at your worksites. You must have labels and signs to warn people about these chemicals. You must have material safety data sheets for all hazardous chemicals. MSDSs provide information from the manufacturer about what is in a chemical and what are its health effects. In addition, employees must be trained about the chemicals to which they are exposed. The training must include information about what you are doing to protect employees and what they need to do to protect themselves. (See 29 CFR 1910.1200.)

Individual Employee Health Records

Employee visits to the health office will generate records that should be kept in order to document what treatment is being provided and how the employee responded. These records help improve the consistency of the health care. They can also be analyzed to discover possible unrecognized hazards, as described earlier in the section about analysis of employee OHDS visits.

QUALIFICATIONS OF OHDS PERSONNEL

Once you have decided which safety and health services you want to provide through your OHDS, you need to decide who will provide the services. There are several factors to consider in making this decision.

First, your OHDS must be organized so that the personnel providing the services are not working alone when state law requires that they be supervised by a registered nurse or physician. Second, the occupational health professionals you use must have specialized, up-to-date training or experience in the methods of occupational health care. Finally, you must choose whether to hire OHDS personnel as your own employees or whether to contract for their services outside your company.

Occupational Health Professionals

Occupational health professionals are medical doctors (MDs), doctors of osteopathy (DOs), and registered nurses (RNs). They hold a license to practice their professions and they practice independently under this license, using the type of standard procedures described in this chapter.

Occupational Medicine Specialists

Occupational medicine specialists are MDs or DOs who have additional training or experience in the treatment of work-related illnesses and injuries. This training or experience can be acquired in different ways. A number of universities offer residency programs that will enable MDs and DOs to become board certified in occupational medicine through the American Board of Preventative Medicine. Professionals graduating from medical school or osteopathic training before January 1, 1984, may qualify for board certification by the "alternative pathway," which requires a combination of education and experience practicing in occupational medicine.

There are shorter programs, called mini-residencies, which offer academic training in occupational medicine. In addition, MDs and DOs can obtain advanced education in occupational medicine by taking continuing education courses. There are MDs and DOs working in occupational medicine who have gained advanced knowledge through experience working in the field for extended periods. A qualified occupational medicine specialist should be capable of performing or managing all the OHDS activities described in the previous section.

Occupational Health Nurses and Nurse Practitioners

Occupational health nurses are RNs who have received specialized training in occupational health. Like occupational medicine physicians, occupational health nurses may gain this training in formal college programs, through continuing education, or through experience working in the field. Professional registered nursing licensure (RN) allows nurses to perform many health evaluation and care functions independently. An occupational health nurse should be capable of performing or managing most of the OHDS activities described in the previous section.

Nurse-practitioners are registered nurses who have completed formal advanced training in physical assessment and the management of minor, stable illnesses and injuries. In some states, they are certificated for advanced practice by state licensing boards. Nurse-practitioners perform many health evaluation and health care activities independently. They may perform many other such activities, under the direction of a physician, through the use of the type of standard procedures known as protocols in the medical community. The physician is not required to be physically present. Nurse-practitioners can also take additional training and specialize in the treatment of occupational illnesses and injuries. When working with standard procedures, nurse-practitioners should be capable of performing all the OHDS program activities described above.

Support Personnel

Support personnel can provide more limited services. They have received specific training and are usually certificated or licensed by the educational institution where they received the training. Sometimes licenses or certification are granted by the state. The scope of practice for support personnel requires that they work under the supervision of licensed health professionals except when delivering first aid. Licensed vocational nurses (LVNs), licensed practical nurses (LPNs), emergency response personnel, and first aid personnel are in this category.

Licensed Vocational or Licensed Practical Nurses

LVNS and LPNs are licensed by state agencies to perform certain health care activities such as taking blood pressures and applying dressings. They must practice under the supervision of a physician or a registered nurse.

First Aid and Emergency Response Personnel

A person does not need a formal health care educational background to be trained in first aid and cardiopulmonary resuscitation (CPR). A certification in first aid is usually granted by the training provider offering the training, after the student completes a standard curriculum and demonstrates competence. The American Red Cross is one example of an institution offering this type of instruction. "First aiders" perform treatments such as splinting and application of ice or pressure dressings. They also learn how to transport ill or injured people. First aiders fulfill the role of first responders. They can provide temporary treatment until care of the ill or injured person is transferred to someone with more advanced health care training. When employers use their employees as first aiders, the employees should be properly trained and certificated because even first aid techniques can be harmful when done incorrectly. The training should be updated on a regular basis.

The teaching of cardiopulmonary resuscitation (CPR) is sometimes included in first aid courses but may have to be taken separately. CPR is the act of providing temporary life-sustaining artificial circulation and respiration when these functions have stopped, before and while the person is transferred to a medical facility. As with first aid, persons completing CPR training are certificated by the organization offering the training. CPR performed incorrectly can injure the person receiving it. When employers use employees as first aiders and expect them to perform CPR, the employees must be thoroughly trained, and this training must be updated at least annually.

Note: It is possible for persons rendering first aid or CPR to be exposed to the viruses that cause Hepatitis B and Acquired Immune Deficiency Syndrome (AIDS), because the first aider can come in contact with body fluids, such as blood, that might contain these viruses. It is crucial that all first aid courses given to employees to qualify them to participate on company emergency teams include information about how the employees can protect themselves from potentially infectious body fluids while performing first aid or CPR.

Emergency response technicians (EMTs) have received advanced training in first aid, CPR, and other life-support techniques. Within certain restrictions they can perform sophisticated emergency procedures and transport ill and/or injured people.

Records of the original training, licenses, update courses, and certification of all employees participating in the delivery of occupational health services, including first aid, CPR, and/or emergency response activities, should be retained within the OHDS.There are numerous types of health care providers qualified to perform services within your OHDS. When considering the selection of an occupational health professional, it is important for you to make sure that their licenses are current and that they have received recent advanced training in occupational health care either through a residency or education program or through experience with enough time in the field to have advanced knowledge.

When considering how to use support personnel, it is important for you to make sure that their licenses or certificates are up to date. It is also important to make sure that they are practicing within their scope of practice and under proper supervision.

Evaluating the Qualifications of a Health Care Professional

Whether you choose to employ professional health care personnel to develop and/or manage your OHDS or choose to contract with outside vendors, it is important to evaluate their qualifications. Remember that occupational health professionals are people selling a service. You should use the same smart-purchaser skills with physicians and registered nurses as you use with anyone who is trying to sell you something. This means knowing something about the services that you want to purchase.

Following are some questions to ask prospective professional health care providers.

1. What type of training has the health care provider had?

A. Physicians:

- Graduation date(s); note all degrees
- Request a copy of current MD or DO license
- Types of specialization certification
- Years of experience in occupational medicine
- Titles of continuing education courses taken in the last two years
- History of professional liability insurance claims against the physician

B. Registered Nurses:

- Graduation date and type of degree
- Graduate degrees
- Types of specialization certification
- Years of experience in occupational health
- Titles of continuing education courses taken in the last two years
- Copy of current RN license

2. In what type of industries has the prospective health care provider had experience?

3. What kind of information does the prospective

health care provider want to know about your business? A prospective health care provider should ask questions about the following:

- Your work processes
- Your known or potential hazards
- Type and location of your facilities
- Number of employees
- Standards and/or regulations that apply in your business
- Health surveillance programs that have been done previously or are in progress
- Collective bargaining contracts
- Any OSHA citations previously issued
- Existence of a safety and health policy and its content
- Current method of providing OHDS services
- Other health care providers providing services

4. What does this provider know about OSHA recordkeeping requirements?

5. Given the above information, what would this provider do to contribute to the improvement of your safety and health program?

6. Can this provider provide references?

   - Has there ever been an OSHA inspection in a facility with which this provider was associated?
   - What was the outcome of that inspection?

Protocols (Established Standardized Procedures)

Protocols are documents that describe your standardized procedures—how you want something done in your company. You may already have standardized procedures for things like your accounting system or for servicing your equipment. Standardized procedures should be written for the activities of your OHDS. In your OHDS, they produce consistency in providing care for work-related illness and injury, for reacting to emergency situations, and for collecting data from health surveillance programs and all the other activities of your OHDS. Standardized procedures are not meant to interfere with an occupational health professional's treatment of work-related injury and illness; they are aimed at ensuring the early detection of work related health problems through consistent and thorough evaluation of employee health complaints.

Standardized procedures also promote the treatment of work-related illness and injury using the most up-to-date therapies. They are particularly important if you are using several contractors to provide your company's OHDS services because they will help ensure that all your employees are receiving the same type of care. Even if you have company employees providing OHDS services, there should be standardized procedures written for all health surveillance programs, health care, and first aid given through your OHDS. These standardized procedures should be communicated to all health care employees and subcontractors providing treatment for your workers.

The American Association of Occupational Health Nurses (AAOHN) has published *A Comprehensive Guide for Establishing an Occupational Health Service,* which includes information on developing protocols. To obtain a copy of this guide, contact AAOHN, 50 Lenox Points, Atlanta, Georgia 30324.

Summary

Your business can use an OHDS to deliver a number of important services. As an employer, you need to consider the special characteristics of your business in order to determine which of these services are appropriate. The factors involved in making this decision include the type of processes and materials your employees work with and the resulting or potential hazards. Other things to think about are the type of facilities in which they work, the number of employees at each site, and the characteristics of this work force, such as age, gender, cultural background, and educational level. The location of each operation and its nearness to a health care facility is also important.

12

Safety and Health Training

Can your employees explain every hazard to which they are exposed or potentially exposed? Can they explain how to protect themselves and their co-workers from these hazards? Can they tell you precisely what they are to do in case of a fire or other emergency? If the answer to any of these questions is no, you may not be training your employees effectively.

Safety and health training is the fourth major element in the Safety and Health Management Guidelines for establishing an effective program for the management of worker safety and health protection. The goal of this training is to help the employees understand safety and health hazards, and how to protect themselves and others from such hazards. It is most effective when it is incorporated into other training about performance requirements and job practices.

Most employers present some form of occupational safety and health training, even if it consists of the explanation of precautions to take when showing a new worker how to do his or her job. The question is, how effective is the training that you do for safety and health? This chapter will help you design (or revise), implement, and evaluate the worker safety and health training that you provide to your employees. It also offers information on OSHA requirements for training and tells you where to find further OSHA references.

Design

Pretraining Elements

Before training begins, be sure that your company policy clearly states the company's commitment to health and safety, including the training program. This commitment must include paid work time for training, and training in the language that the worker understands. Involve both management and employees in the development of the training. Be sure that health and safety concerns can be improved by training, and that specific goals for the training are clearly identified.

Teaching and Learning Principles

Training your supervisors and employees need not be complex or lengthy. In most small businesses, the owner or manager can do the training. In larger businesses with extensive training needs, more formal training may be necessary. In either case, five basic principles should guide your training program.

1. *Perceived Purpose*: The learner must understand the purpose for the learning. Therefore, the beginning of any training program should focus on why this instruction will be useful to the learner.

2. *Order of Presentation*: Training should be presented so that the organization and meaning are clear to the learners. For example, if you are teaching employees the proper way to use a respirator, the order in which you present the material should match the steps the employee must use to choose, fit, wear, and maintain the respirator.

3. *Appropriate Practice*: We learn best when we can immediately practice and apply newly acquired knowledge and skills. Therefore, instruction for job safety and health is best given at the worksite where demonstration, practice, and application can be immediate. When on-site instruction is not feasible, plan for practice and application of the new knowledge and skills as soon as possible.

4. *Knowledge of Results*: As we practice, we need to know as soon as possible if we are correct. Incorrect practice can establish harmful patterns, while praise for correct actions will enhance motivation as well as begin to establish the desired results.

5. *Individual Differentiation*: Since we are individuals, we learn in different ways. A successful training program incorporates a variety of learning opportunities, such as written instruction, audio-visual instruction, lectures, and coaching in hands-on situations. Also, we learn at different speeds, and we learn through teaching others. Therefore, after the initial instruction and some practice, dividing the group into teaching and learning pairs, pairing a slower learner with a quicker one is often helpful.

Identifying Training Needs

New employees need to be trained not only to do the job for which they were hired, but also to recognize, understand, and avoid the potential hazards to themselves and others during the performance of that job and elsewhere within the workplace. Contract workers may also need training to recognize the hazards or potential hazards in your workplace. Experienced workers will need training if the installation of new equipment changes their job in any way, or if process changes bring new hazards into the workplace or increase hazards already there. All workers may need refresher training to keep them alert to ongoing housekeeping problems and to keep them prepared for emergencies. Workers needing to wear personal protection equipment and persons working in high-risk situations will need special training.

Specific hazards that employees need to know about should be identified through total site health and safety surveys, job hazard analysis, and change analysis. Company accident and injury records may reveal other hazards and needs for training. Near-miss reports, maintenance requests, and employee suggestions may uncover other hazards for which employees may need additional training to help them recognize these hazards and to know how to protect themselves from them.

As you start or revise your program to prevent or control those hazards, you will probably use some controls that require the cooperation of your workers, such as wearing personal protective equipment (PPE) properly or carrying out certain parts of their jobs with precautions. Your workers will need training to be sure that they understand the actions you want them to take and know why the actions are important. Looking at all of these things, you will be able to identify what training is needed.

Developing Learning Activities

Keeping in mind the five learning principles described above, develop your learning activities to meet the training needs you have identified. Be imaginative in your choice of methods and materials. Make full use of your resources. You can get ideas about training by looking at similar training programs offered by other companies in the same industry, or by obtaining suggestions from such organizations as the National Safety Council, the American Society of Safety Engineers, the Bureau of Labor Statistics, OSHA consultation programs, and the OSHA Office of Training and Education.

Some Usual Types of Specialized Training

Safety and Health Training for Managers

A good safety and health program is impossible without support and understanding from the top. Training for managers in their responsibilities is necessary to ensure that continuing support and understanding. Formal classroom training may not be necessary. The subject can be covered periodically as a part of regular management meetings.

Managers should understand the importance of the goal and objectives set for safety and health and clearly communicate them to their employees. They should remember to make clear assignments of safety and health responsibilities, provide authority and resources to carry them out, and hold subordinate managers and supervisors accountable. It is important that they let employees see their support for the safety and health program and that they set a good example by scrupulously following all the safety and health rules. They should also actively encourage employee involvement in safety and health problem identification and resolution. These topics can be covered and illustrated with examples in a relatively short time. They should be repeated at least annually.

Safety and Health Training for Supervisors

Although all employees should be involved in matters of safety and health, workers are often promoted to supervisory positions without adequate knowledge of how to train other employees in the safe and proper way to do the job and without full knowledge of the company's policies and procedures. They may need additional training in detecting and controlling hazards, investigating accidents, their role in ensuring the maintenance of physical protections, handling emergencies, and, in general, how to supervise.

Since supervisors do a lot of on-the-job training, they will also need training in how to train and how to reinforce training. They will also need help in learning how to apply fair and consistent discipline, when necessary. Such training can be provided by the supervisor's immediate management, by the Safety Department, or by outside sources such as the OSHA Consultation Service. The OSHA Training Institute is one good source for some of the training.

Job Orientation

The format and extensiveness of orientation training will depend upon the complexity of hazards and the work practices needed to control them. For many small businesses, job orientation may consist of a quick review of site safety and health rules, hazard communication training for the toxic substances present at the worksite, and a run-through of the tasks of the job. This training is usually presented by the new employee's supervisor or by a combination of the personnel officer and the supervisor. Even for this simplified version of orientation, the supervisor should make sure that the new employee also receives information about response to emergencies before the new job is begun.

For larger workplaces with more complex hazards and work practices to control them, orientation should be carefully structured to ensure that the new employee starts the job with a clear understanding of the hazards and how to protect himself or herself and others. This type of training is frequently a combination of classroom and on-the-job training. Many companies have found it useful to train fellow employees to present the training to their peers. Others have followed up supervisory training with a "buddy system" in which a worker with longtime experience is assigned to watch over and coach a new worker for a set period of time or until the "buddy" and supervisor determine that training is complete.

Vehicular Safety

In 1987, the leading cause of work-related deaths was over-the-road motor vehicle accidents. Therefore, all workers operating a motor vehicle as part of their job should be trained in its safe operation. In 1990, OSHA proposed a standard requiring that such workers wear seat belts and receive safe driving training. Training in safe loading and unloading practices, safe speed in relation to varying conditions, and proper vehicle maintenance has been found to be helpful in reducing work-related vehicle injuries. The benefits of safe driving and clear consequences for unsafe practices should be communicated to all drivers.

Similar training should be given to drivers of on-premises vehicles. Good employers train and license such drivers because of the special hazards to which they can be exposed. These hazards include vehicle balance, cornering where loads can tip, and the special precautions related to battery charging.

Personal Protection Equipment (PPE)

Both supervisors and workers must be taught the proper selection, use, and maintenance of PPE. Since PPE can sometimes be cumbersome, employees may need to be motivated to wear it in every situation where protection is necessary. Therefore, training should begin with a clear explanation as to why the equipment is necessary, how its use will benefit the wearer, and what its limitations are. Remind your employees of your desire to protect them and of your efforts, not only to eliminate and reduce the hazards, but also to provide suitable personal protection equipment where it is needed. Explain how necessary it is that they do their part to protect their health and safety.

It is important that individual employees become familiar with the PPE they are being asked to wear by handling it and putting it on. Show them how to put it on and wear it properly. Show them any test that they should make as users to ensure proper fit. Proper fit is necessary to obtain the protection the equipment was designed to provide, and is especially important in the case of negative pressure respirators, so special fit testing will be necessary.

In addition to fit and use, some PPE requires proper maintenance in order to function effectively. Employees must be trained to maintain the equipment or to see that it is maintained properly. Vendors of the equipment and manufacturers' instructions are often your best source of this information.

If emergency procedure plans call for employees who do not regularly use PPE to use it as part of the emergency action, they also must be trained in the use, fit, and maintenance of the equipment. Simulated emergency training exercises in which employees use the equipment should be part of your overall training program. Repeated training is often necessary for new hires, contract workers, employees in new jobs, and for emergency use of PPE. For example, training on the use of self-contained respirators used to escape from atmospheres immediately dangerous to life or health should be repeated frequently enough so the you can be certain that each employee who depends on these respirators will lose no time finding and donning the protective device when it is needed.

Emergency Response

At every work site employees should be trained to respond to emergency situations. The training should include making sure that each employee has a clear understanding of emergency telephone numbers and who may use them, of emergency exits and how they are marked, of evacuation routes, and of the signals that alert employees to the need to evacuate. In addition, evacuation drills should be practiced at least annually so that every employee has a chance to recognize the signal and evacuate in an orderly fashion. Supervisors or their alternates should practice counting personnel at evacuation gathering points to ensure that every employee is accounted for.

Do not forget any workers at your site when you are practicing for emergencies. You should have procedures for accounting for visiting vendors, contract employees, and service workers such as cafeteria employees. All workers at the site should be able to hear the signal and practice the needed emergency drills. It may be helpful for you to speak quietly to one or two employees about simulating an emergency-caused injury or other problem that would keep them from evacuating the building. They can do that by slipping away to a forgotten corner, such as a stairwell, that is not part of the evacuation route and await "discovery." That will give your supervisors or alternates a better understanding of the importance of an accurate count.

At work sites where weather or earthquake emergencies are possible occurrences, additional special instruction and drilling may be necessary. For example, where there is a good possibility of tornado emergencies, employees should learn to differentiate the signals for evacuation and for tornado shelter and practice the correct response.

Periodic Safety and Health Training

At worksites where complex work practices are necessary to control hazards or where occupational injuries and illness are prevalent, employees will need periodic safety and health training to refresh their memories and to teach new methods of control. Also, new training may be necessary to instruct employees when OSHA standards change or new standards are promulgated. It is important to keep these sessions interesting. Some worksites have found that giving employees the responsibility to plan and present periodic safety and health training is very effective. This method can only be effective, however, if employees are provided with training resources to use in the development of their presentations.

Most general industry worksites use monthly safety meetings for this training. In construction and other rapid-change, high-hazard industries, weekly meetings are conducted to remind workers of the tasks they will take on during the course of the week, the environmental changes that may affect them, and what procedures will be needed to protect themselves and others.

Perhaps the most effective training can be done by the supervisor with each individual employee periodically. This is called one-on-one training. The supervisor spends some time watching the employee work. Then he or she meets with the employee to discuss safe work practices, giving credit to the employee for safe work and providing additional instructions for tasks in which unsafe practices were observed. It is important to use this method with all employees under supervision and not just those with whom there appears to be a problem. *Note*: Providing positive feedback for successful use of safe work practices is the most powerful training tool in this system.

Conducting Training

Employees must be motivated enough to want to learn and to improve. You can enhance the success of the training in the following ways:

1. Prepare employees for training by putting them at ease.
2. Recruit employees who would be good trainers of their peers and prepare them to do the training.
3. Explain the job or training topic and determine how much your employees already know about it.
4. Boost their interest in learning about the job by helping them see how the training will help them.
5. Pace the instruction to the speed of the learners, presenting it clearly and being patient.
6. Present no more information in one session than your employees can master.
7. Have your employees perform each step of the operation and repeat your instructions as to why it is done the way you describe. Have them repeat a sequence until you are satisfied that they know how to do it.
8. Encourage employees to help each other by dividing the group into teacher–learner pairs or practice pairs.
9. During the initial practice period, check on the employees frequently for correct performance. Taper off on the checking as they become more proficient.
10. Encourage your employees to incorporate the new skill into the way they work best, but caution them not to change the newly learned procedure unless they first check with you or their supervisor.

Evaluation

A plan for evaluating the training session(s) should be generated as the needs are being identified and the content is being developed, not delayed until the training is completed. Evaluation will help employers and supervisors ensure that the desired learning has been achieved and that the employee's performance has improved on the job. These are some methods of evaluating training:

1. Ask employees' opinions—through questionnaires or during informal discussions—to assess the relevance and appropriateness of the training program.

2. Test the results of the training by stopping at the work station to observe an employee's performance both before and after training, noting improvement or changes. After the training you can measure its success a second time by asking the worker for an explanation of the hazards involved in the work and how the worker can protect himself or herself. If the training has been well presented and understood, each trained worker should be able to give a good, comprehensive response and show improvement or change.

3. Record workplace improvements, such as reduced injury or accident rates.

However it is conducted, an evaluation of training can give employers the information necessary to decide whether or not the employees achieved the desired results, whether the training session should be offered again at some future date, and if any modifications need to be made in the training program.

Recordkeeping

Even if you operate a very small business, it is to your advantage to keep training records. A simple format is all that is needed: who was trained, on what topic or job, on what date, and an evaluation of the learner's participation and success. These records will help you make sure that everyone who needs training will receive it, that refresher courses will be provided at regular intervals, and that documentation will be available, if necessary, to show that training was done appropriately. On the following page is a sample form for keeping such records.

Sources for Assistance

In addition to this book, assistance with training programs or the identification of resources for training is available through such organizations as OSHA full-service area offices, state agencies with their own OSHA-approved occupational safety and health programs, OSHA-funded state on-site consultation programs for employers, local safety councils, the OSHA Office of Training and Education, and OSHA-funded grantees.

Summary

Good safety and health training programs will benefit the employee, the employer, and the company as a whole because such programs demonstrate an honest concern for health and safety. They help improve workers' morale and often decrease near-miss incidents and accidents, resulting in fewer lost work days and more production. Records of such training provide evidence that employees have

EMPLOYEE TRAINING RECORD

Employee Name: ___________________________________

Employee Number: _________________________________

Department: _____________________________________

Occupation(s): ___________________________________

JOB	DATE TRAINED	DATE RETRAIN	COMMENTS

I have received and understood the safety and health training and/or reinstruction listed above.

_________________________ _________________________

Employee Signature Supervisor Signature

been trained and can be used to determine what training employees still need to receive.

Safety and health training programs consist of determining training needs and involving employees in the program design and in the teaching. New hires, contract workers and workers, in high risk areas should be especially targeted for training. Use the five principles of teaching and learning when constructing course content. Evaluate each course and modify when necessary. Always keep orderly records, noting not only who was trained, but also that the trainee learned what was taught.

Appendix 1: Preparing a Policy Statement Worksheet

Policy statements can vary in length and content, depending on whether they contain only policy or also include company philosophies, rules, and procedures. Some policy statements, for instance, will cover items such as specific assignment of responsibility, delegation of authority, description of duties, safety and health rules, and provision for a safety committee. While some companies may wish to include these additional items in the policy statement, it is usually best to leave the details for later discussion.

This worksheet is designed to help you develop your safety and health policy statement. It contains examples of specific statements often found in safety and health policies. They are only examples, but from them you may get ideas on how to write a policy statement that expresses your style, your attitudes, and your values about the safety and health of your employees.

Introductory Statement

The written policy statement generally starts with a clear, simple expression of your concern for and attitude about employee safety and health. Examples of introductions to policy statements include:

- This company considers no phase of its operation or administration more important than safety and health. We will provide and maintain safe and healthful working conditions, and establish and insist on safe work methods and practices at all times.
- Accident prevention is a primary job of management, and management is responsible for establishing safe and healthful working conditions.
- This company has always believed that our employees are our most important asset. We will always place a high priority on safe operations and on the safety and health of employees.
- People are our most important resource and their safety and health are among our principal responsibilities.
- The company will, at all times, and at every level of management, attempt to provide and maintain a safe and healthful working environment for all employees. All safety and health protection programs are aimed at preventing accidents and exposures to harmful atmospheric contaminants.
- All members of management and all employees must make safety and health protection a part of their daily and hourly concern.

Purpose and Philosophy

An effective safety and health program will have a stated purpose or philosophy. These are included in the written policy statement so that both you and your employees are reminded of the purpose and value of the safety and health program. You could incorporate into your policy such statements as:

- We have established our safety and health program to eliminate employee work-related injuries and illnesses. We expect it to improve operations and reduce personal and financial losses.
- Safety and health protection shall be an integral part of all operations, including planning, development, production, administration, sales, and transportation.
- We want to make our safety and health protection efforts so successful that we make elimination of accidents, injuries, and illnesses a way of life.
- We aim to resolve safety and health problems through prevention.
- We will involve both management and employees in planning, development, and implementation of safety and health protection.

Management Responsibilities

Your safety and health action plan will describe in detail who is to develop the program and make it work, as well as the specific assignment of responsibilities, duties, and delegation of authority. The policy statement may summarize these responsibilities, as illustrated below:

- The company management is responsible for developing an effective safety and health program.
- The plant superintendents are responsible for maintaining safe and healthful working conditions and practices in the areas under their jurisdiction.

- Department heads and supervisors are responsible for preventing accidents and health hazard exposures in their departments.
- Foremen are responsible for preventing accidents and health hazard exposures to their crews.
- Supervisors will be accountable for the safety and health of all employees working under their supervision.
- The Safety Director has the authority and responsibility to provide guidance and to aid supervisors in preventing accidents and health hazard exposures.
- Each level of management must reflect an interest in company safety and health and set a good example by complying with company rules for safety and health protection. Management interest must be vocal, visible, and continuous from top management to departmental supervisors.
- Management representatives who have been assigned safety and health responsibilities will be held accountable for meeting those responsibilities.

Employee Responsibilities

Many companies acknowledge the vital role of their employees in the operation of a successful safety and health program by summarizing their roles and contributions in the policy statement. Following are some examples:

- All employees are expected to follow safe working practices, to obey rules and regulations, and to work in a way that maintains the high safety and health standards developed and sanctioned by the company.

- All employees are expected to give full support to safety and health protection activities.
- Every employee must observe established safety and health regulations and practices, including the use of personal protective equipment.
- All employees are expected to take an active interest and participate in the safety and health program, and abide by the rules and regulations of this company.
- All employees must recognize their responsibility to prevent injuries and illnesses and take necessary actions to do so. Their performance in this regard will be measured along with their overall performance.

Closing Statement

The closing statement may reaffirm your commitment to provide a safe and healthful workplace, and may be an appeal for the cooperation of all company employees in support of the safety and health program.

- I urge all employees to make this safety and health program an integral part of their daily operations.
- By accepting mutual responsibility to operate safely, we will all contribute to the well-being of personnel and subsequently the company.
- We must be so successful in our efforts that total elimination of accidents, injuries, and illnesses becomes a way of life.

Summary

Generally, a Safety and Health Policy Statement should be between six and twelve sentences long. It should include the five elements listed above: an introductory statement, a statement of the purpose or philosophy of the policy, a summary of management responsibilities, a summary of employee responsibilities, and a closing statement.

Following is one example of a Safety and Health Policy Statement:

> This company considers no phase of its operation more important than safety and health protection. We will provide and maintain safe and healthful working conditions and establish and insist upon safe work methods and practices at all times.
>
> Safety and health shall be an integral part of all operations including planning, development, production, administration, sales, and transportation. Accidents have no place in our company.
>
> We will work consistently to maintain safe and healthful working conditions, to adhere to proper operating practices and procedures designed to prevent injury and illness, and to observe federal, state, local, and company safety and health regulations.
>
> Each level of management must reflect an interest in company safety and health objectives and is required to set a good example by always observing the rules as a part of their normal work routine. Management interest must be vocal, visible, and continuous, from top management to departmental supervisors.
>
> All employees are expected to follow safe working practices, to obey rules and regulations, and to work in a way that maintains the high safety and health standards developed and sanctioned by the company.
>
> We urge all employees to make our safety and health program an integral part of their daily operations. That way the total elimination of accidents and injuries will become not just an objective, but a way of life.

Appendix 2: Safety and Health Program Responsibilities Worksheet

All employees will be fully responsible for implementing the provisions of our safety and health policy as it pertains to operations under their jurisdiction. The responsibilities listed below are our minimum expectations. Individual initiatives to curb losses are encouraged.

JOB
TITLE:____________________________________

GENERAL STATEMENT:________________________

LIMITS OF AUTHORITY AND RESOURCES (Cost, reporting, and authority to close down equipment):

SPECIFICALLY, THE EMPLOYEE WILL BE RESPONSIBLE FOR AND HELD ACCOUNTABLE FOR:

Appendix 3: Description of a Safety and Health Program Assessment

This description, modeled after the outline in Chapter 7, explains each of the assessment tools and how to use them to assess each element and subelement of a safety and health program.

There are three basic methods for assessing safety and health program effectiveness:

- Checking documentation of activity
- Interviewing employees at all levels for knowledge, awareness, and perceptions
- Reviewing site conditions and, where hazards are found, tracing the root causes of those hazards

Some elements of program management better lend themselves to one method of assessment than do others. Some elements should be assessed using two or all three of the methods.

Documentation

Checking documentation is a standard audit technique. It is particularly useful for understanding whether or not the tracking of hazards to correction is effective, but it can also be used to discover the quality of certain activities such as self-inspections or routine hazard analysis.

Where documentation is done, inspection records can tell the assessor or auditor whether serious hazards are being found or whether the same hazards are found repeatedly. In the first case, if serious hazards are not being found and yet accidents keep occurring, there may be a need to train those who inspect to look for different hazards than they do now.

In the second case, if the same hazards are being found repeatedly, the problem may be more complicated. Perhaps the hazards are not being corrected. That would be a tracking problem or a problem with accountability for hazard correction.

If the same hazards occur again after they are corrected, someone is not taking responsibility for keeping those hazards under control. Either the responsibility is not clear or those who are responsible are not being held accountable.

Employee Interviews

Talking to randomly selected employees at all levels will provide a good indication of the quality of employee training and of employee perceptions of how well the program is running. If the training provided is effective, the employees interviewed can tell about the hazards they work with and how they help to protect themselves and others by keeping those hazards controlled. Every employee should also be able to say precisely what he or she is expected to do, and where to go if an emergency occurs.

Employee interviews can tell about other aspects of your program as well. The employee's perception of how easy it is to report a hazard, and get a timely and appropriate response, will tell a lot about how well your system for employee reports of hazards is working. Employee perception of your system for enforcing safety and health rules and safe work practices can also be enlightening. If they perceive inconsistency or confusion, you will know that the system needs improvement.

Employee interviews are not limited to hourly employees. A lot may be learned from talking with first line supervisors. It is also helpful to find out from some line managers what their perceptions of their safety and health responsibilities are.

Site Conditions and Root Causes of Hazards

Looking at the conditions of the workplace will reveal the hazards that exist there and can also provide much information about the breakdown of management systems meant to prevent or control those hazards.

Looking at conditions and practices is well understood as a technique of assessing the effectiveness of safety and health programs. If, for example, in areas where personal protective equipment is required, you see large and understandable signs designating areas where PPE is required and employees in those areas wearing equipment properly, with no exceptions, you have seen a valuable clue about how well the PPE program is working.

Beyond noting good conditions and practices, you can

also obtain good information about management from the hazards seen through *root analysis*. This approach to hazards is much like the most sophisticated accident investigation techniques in which many contributing factors are located and corrected or controlled.

If, for example, during a review of conditions, you find a machine being operated without a guard on a pinch point, you should not stop with just getting the guard put on. Asking a few questions will tell a lot about one or more widely varied management systems. If the operator says he did not know the guard was supposed to be on the machine, that should lead to a series of questions about the existence of safe work procedures and/or training.

If he says that the guard slows him down and the supervisor knows he takes it off, many questions can be asked about enforcement of rules, accountability, and the clarity of the worksite objective of putting safety and health first.

If, on the other hand, the need for the guard is noticed by your insurance inspector or an OSHA inspector, or you do not notice it until someone is hurt, the questions to be asked should start with whether a comprehensive survey of the worksite has been done by someone with enough expertise to recognize all the potential and existing hazards.

Analyzing the root causes of hazards is a technique that should not wait for formal assessments of your program—it should be in use every day. However, it is also very helpful when used as part of a formal assessment.

To assess each part of your worksite's safety and health program, you should use one or more of these methods, as appropriate. The remainder of this appendix will identify the subelements found in each element of a quality safety and health program and identify methods that may be useful in assessing the subelement.

I. Assessing the Key Subelements of Management Commitment

A. Worksite policy on safe and healthful working conditions

1. Documentation.

a. If there is a written policy, does it make clear the priority of worker safety and health in relation to other organizational values, such as production?

2. Interviews

a. When asked, can employees at all levels give you a fair idea of what your worksite policy on worker safety and health is?
b. If the policy is written, can hourly employees tell you where they have seen it?
c. Can employees at all levels explain the priority of worker safety and health in relation to other organizational values, as the policy intends?

3. Site conditions and root causes of hazards

a. Have injuries occurred because employees at any level did not understand the importance of safety precautions in relation to other organizational values, such as production?

B. Goal and objectives for worker safety and health

1. Documentation

a. If there is a written goal for your safety and health program, is it updated annually?

b. If there are written objectives, such as an annual plan to reach that goal, are they clearly stated?

c. If managers and supervisors have written objectives, do they include objectives for the safety and health program?

2. Interviews

a. Do managerial and supervisory employees have a clear idea of their objectives for worker safety and health?

b. Do hourly employees understand the current objectives of the safety and health program?

3. Site conditions and root causes of hazards: only helpful in a general sense

C. Visible top management involvement

1. Documentation

a. Are there one or more written programs that provide for the involvement of top-level management in safety and health activities? For example, top management can receive and sign off on inspection reports received either after each inspection or in a quarterly summary. These reports can then be posted for employees to see. Top management can provide "open door" times each week or each month for employees to come in to discuss safety and health concerns. Top management can reward the best safety suggestions each month or at other specified intervals.

2. Interviews

   a. Can hourly employees describe how management officials are involved in safety and health activities?
   b. Do hourly employees perceive that managers and supervisors follow safety and health rules and work practices, such as wearing appropriate personal protective equipment?

3. Site conditions and root causes of hazards

   a. When employees are found not wearing required personal protective equipment or not following safe work practices, have any of them said that managers or supervisors also did not follow these rules?

D. Employee involvement

1. Documentation

   a. Are there one or more written programs that provide for employee involvement in decisions that affect their safety and health?
   b. Is there documentation of those activities (such as employee inspection reports, minutes of joint employee-management, or employee committee meetings)?
   c. Is there written documentation of any management response to employee safety and health program activities?
   d. Does the documentation indicate real substance to employee activities?

e. Are there written guarantees of protection for employees from possible harassment resulting from safety and health program activities?

2. Interviews

a. Are employees aware of the way(s) in which they can be involved in decisions that affect their safety and health?
b. Do employees appear to take pride in the achievements of the worksite safety and health program?
c. Are employees comfortable answering questions about safety and health programs and conditions at the site?
d. Do employees feel that they have the support of management for their safety and health activities?

3. Site conditions and root causes of hazards: Not applicable

E. Assignment of responsibility so that everyone knows what performance is expected

1. Documentation

a. Are responsibilities written out so that they can be clearly understood?

2. Interviews

a. Do employees understand their own responsibilities and those of others?

3. Site conditions and root causes of hazards

a. Was the hazard caused in part because

no one was assigned the responsibility to control or prevent it?

b. Was the hazard allowed to exist in part because someone in management did not have the clear responsibility to hold a lower-level manager or supervisor accountable for carrying out assigned responsibilities?

F. Adequate authority and resources so that responsibilities can be met

1. Documentation (only generally applicable)

2. Interviews

a. Do safety staff members or any other personnel have responsibilities to ensure safe operation of production equipment without the authority to shut down that equipment or order maintenance or parts?
b. Do employees talk about not being able to get safety or health improvements because of cost?
c. Do employees mention the need for more safety or health personnel or expert consultants?

3. Site conditions and root causes of hazards

a. Do recognized hazards go uncorrected because of lack of authority or resources to correct them?
b. Are there unrecognized hazards because better expertise is needed to diagnose hazards?

G. Holding managers, supervisors, and hourly employees accountable for meeting their safety and health responsibilities

1. Documentation

   a. Do all line manager and supervisory performance evaluations include specific safety and health protection criteria?
   b. Is there documented evidence of employees at all levels being held accountable for safety and health responsibilities, including safe work practices, either through performance evaluations impacting pay and/or promotions or disciplinary actions?

2. Interviews

   a. When you ask employees what happens to people who break safety and health rules or safe work practices, do they indicate that such employees are held accountable in a clear and consistent manner?
   b. Do hourly employees indicate that supervisors and managers really care about meeting safety and health responsibilities?
   c. When asked about what happens when rules are broken, do hourly employees complain that supervisors and managers do not follow rules and are never disciplined for those infractions?

3. Site conditions and root causes of hazards

   a. Are hazards occurring because employees, supervisors, and/or managers are not being held accountable for their safety and health responsibilities?

b. Are identified hazards not being corrected because those assigned the responsibility are not being held responsible?

H. Evaluate program operations

1. Documentation

a. Is there a written evaluation of each major part of the program that lists what is being done, evaluates how effective it is against the goal and objectives, and recommends changes where needed to make the program more effective or to try other programs instead?

2. Interviews

a. Can employees, supervisors, and/or managers tell you how the whole program is evaluated and revised each year?

3. Site conditions and root causes of hazards: only generally applicable

II. Assessing the Key Subelements of Worksite Analysis

A. Comprehensive surveys, change analysis, routine hazard analysis

1. Documentation

a. Are there documents that provide comprehensive analysis of all potential safety and health hazards of the worksite?

b. Are there documents that provide analysis of potential safety and/or health hazards for each new facility, equipment, material, or process and the means for elimination or control of any hazards identified?
c. Is the step-by-step analysis of the hazards in each part of each job documented so that you can clearly see the analysis that resulted in decisions on safe work procedures?
d. If there are complicated processes with a potential for catastrophic impact from an accident but low probability of an accident—such as in nuclear power or chemical production—are there documents analyzing the potential hazards in each part of the processes, and the means to prevent or control them?
e. If there are processes with a potential for catastrophic impact from an accident but low probability of an accident, have analyses of the processes such as "fault tree" or "what if?" been documented to ensure enough backup systems for worker protection in the event of more than one control failure?

2. Interviews

a. Do employees complain that new facilities, equipment, materials, or processes are hazardous?
b. Do any employees say that they have been involved in job safety analysis or process review and are satisfied with the results?

c. Does the safety and health staff indicate ignorance of existing or potential hazards at the worksite?
d. Does the occupational nurse, doctor, or health care provider understand the potential occupational diseases and health effects in this worksite?

3. Site conditions and root causes of hazards

   a. Have hazards appeared where no one in management realized there was potential for their development?
   b. Where workers have followed job procedures faithfully, have accidents or near-misses occurred because of hidden hazards?
   c. Have hazards been discovered in the design of new facilities, equipment, materials, or processes after use was begun?
   d. Have accidents or near-misses occurred when two or more failures in the hazard control system surprised everyone by occurring at the same time?

B. Regular site safety and health inspections

1. Documentation

   a. If inspection reports are written, do they show that inspections are done on a regular basis?
   b. Do the hazards found indicate good ability to recognize hazards that are typical for this industry?
   c. Are hazards found during inspections tracked to complete correction?
   d. What is the relationship between the

hazards found in inspections and those found to be at least in part responsible for injuries or illness?

2. Interviews

   a. Do employees indicate that they see inspections being done that seem thorough?

3. Site conditions and root causes of hazards

   a. Are the hazards discovered during accident investigations ones that should have been recognized and corrected by the inspection process?

C. Employee reports of hazards

1. Documentation

   a. Is the system for written reports being used frequently?
   b. Are valid hazards that have been reported by employees tracked to complete correction?
   c. Are the responses timely and adequate?

2. Interviews

   a. Do employees know whom to contact and what to do if they see something they believe to be hazardous to themselves or co-workers?
   b. Do employees feel that responses to their reports of hazards are timely and adequate?
   c. Do employees mention that sometimes when they report a hazard they never hear anything further about it?

d. Do any employees mention that they or other employees are being harassed officially or otherwise for reporting hazards?

3. Site conditions and root causes of hazards

a. When hazards are found, is it ever the case that employees say they have been complaining about them for a long time but no one has done anything to achieve correction?
b. Are hazards ever found where you would think employees would have been sure to recognize them but never reported them?

D. Accident and near-miss investigations

1. Documentation

a. Do accident investigation reports show a thorough analysis of the cause(s) of accidents, rather than quickly blaming the injured employee?
b. Are near-misses (property damage or "close calls") investigated using the same techniques as accident investigations?
c. Are hazards identified as contributing to accidents or near-misses tracked to correction?

2. Interviews

a. Do employees understand and accept the results of accident and near-miss investigations?
b. Do employees mention a tendency on

management's part to blame the injured employee?

c. Do employees feel that all hazards contributing to accidents are corrected or controlled?

3. Site conditions and root causes of hazards

a. Are accidents sometimes caused at least in part by factors that might have contributed to near-misses that were not investigated, or to accidents that were too quickly investigated?

E. Injury and illness pattern analysis

1. Documentation

a. In addition to the required OSHA log, are careful records kept of first aid injuries and/or illnesses that might not immediately appear to be work-related?
b. Is there any periodic, written analysis of the patterns of near-misses, injuries, and/or illness over time to see if there are previously unrecognized connections between them that indicate unrecognized hazards needing correction or control?
c. Looking at the OSHA 200 Log and, where applicable, first aid logs, are there patterns of illness or injury that should have been analyzed for previously undetected hazards?
d. If there is an occupational nurse or doctor on the worksite or if employees are encouraged to take ordinary illness to a nearby health care provider, are the lists

of those visits analyzed to see if there are any clusters of illness that might be work-related?

2. Interviews

   a. Have employees mentioned illnesses or injuries that seem related to them but that have not been analyzed for previously undetected hazards?

3. Site conditions and root causes of hazards: not generally applicable

III. Assessing the Key Subelements of Hazard Prevention and Control

A. Appropriate use of engineering controls, work practices, personal protective equipment, and administrative controls

1. Documentation

   a. If there are documented comprehensive surveys, are they accompanied by a plan for systematic prevention or control?
   b. If there is a written plan, does it show that the best method of hazard protection was chosen?
   c. Are there written safe work procedures?
   d. If respirators are used, is there a written respirator program?

2. Interviews

   a. Do employees say that they have been trained in and have ready access to reliable safe work procedures?
   b. Do employees mention that they have

difficulty accomplishing their work because of unwieldy controls meant to protect them?

c. Do employees ever mention personal protective equipment (PPE), work practices, or engineering controls as interfering with their ability to work safety?
d. Do employees who use PPE understand why they use it and how to maintain it?
e. Do employees who use PPE indicate that the rules for PPE use are consistently and fairly enforced?
f. Do employees indicate that safe work procedures are fairly and consistently enforced?

3. Site conditions and root causes of hazards

a. Do you ever find that controls meant to protect workers are actually putting them at risk or not providing enough protection?
b. Are employees involved in unsafe practices or creating unsafe conditions because rules and work practices are not fairly and consistently enforced?
c. Are employees in areas designated for PPE wearing it properly and without exception?
d. Are hazards that could feasibly be controlled by better design being inadequately controlled by other means?

B. Provide facility and equipment preventive maintenance

1. Documentation

a. Is there a preventive maintenance schedule that provides for timely maintenance of the facilities and equipment?
b. Is there a written or computerized record of when maintenance was performed that shows the schedule has been followed?
c. Do maintenance request records show a pattern of certain facilities or equipment needing repair or breaking down before maintenance was scheduled or before it was actually done?
d. Do any accident/incident investigations list facility or equipment breakdown as a major cause?

2. Interviews
   a. Do employees mention difficulty with improperly functioning equipment or facilities in poor repair?
   b. Do maintenance employees feel that the preventative maintenance system is working well?
   c. Do employees feel that hazard controls needing maintenance are properly cared for?
3. Site conditions and root causes of hazards
   a. Is poor maintenance a frequent source of hazards?
   b. Are hazard controls in good working order?
   c. Does equipment appear to be in good working order?

C. Plan and prepare for emergencies

1. Documentation

a. Are there written procedures that clearly cover every likely emergency with clear evacuation routes, assembly points, and emergency telephone numbers?

2. Interviews

a. Can employees tell you exactly what they are supposed to do in case of any kind of likely emergency that you ask them about?

3. Site conditions and root causes of hazards

a. Have hazards occurred in actual or simulated emergencies due to confusion about what to do?
b. In larger worksites, are emergency evacuation routes clearly marked?
c. Are emergency telephone numbers and fire alarms easy to find?

D. Establish a medical program

1. Documentation

a. Are good, clear records kept of medical testing and assistance?

2. Interviews

a. Do employees say that test results were explained to them?
b. Do employees feel that more first aid or CPR-trained personnel should be available?
c. Are employees satisfied with the medical arrangements at or for the site?
d. Does the occupational health care provider understand the potential hazards

of the worksite so that occupational illness symptoms can be recognized?

3. Site conditions and root causes of hazards

   a. Have further injuries or worsening of injuries occurred because proper medical assistance—including trained first aid and CPR—was not readily available?
   b. Have occupational illnesses gone undetected because no one with occupational health specialty training reviewed employee symptoms as part of the medical program?

IV. Assessing Key Subelements of Safety and Health Training

A. Ensuring that all employees understand hazards

1. Documentation

   a. Does the written training program include complete training for every employee in every potential hazard to which each employee may be exposed and in emergency procedures?
   b. Do training records show that every employee received the planned training?
   c. Do the written evaluations of training indicate that the training was successful and that the trained employees learned what was intended?

2. Interviews

   a. Can employees tell you what hazards they are exposed to, why those hazards are a threat, and how they can help protect themselves and others from those hazards?

b. If PPE is used, can employees explain why they use it and how to use and maintain it?

c. Can employees tell you precisely what they are supposed to do and where they are supposed to go in every kind of emergency likely to occur at your worksite?

d. Do employees feel that health and safety training is adequate?

3. Site conditions and root causes of hazards

   a. Have employees been hurt or made ill by hazards of which they were completely unaware, or from which they did not understand the danger, or from which they did not know how to protect themselves?

   b. Have employees or rescue workers ever been endangered by employee lack of knowledge about what to do or where to go in any given type of emergency situation?

   c. Are there hazards in the workplace that exist, at least in part, because one or more employees was not adequately trained in hazard control?

   d. Are there any instances of employees not wearing required PPE properly because they have not been properly trained?

B. Ensuring that supervisors understand their responsibilities

1. Documentation

a. Do training records indicate that all supervisors have been trained in their responsibilities to analyze work under their supervision for unrecognized hazards, to maintain physical protection, and to reinforce employee training through performance feedback and, where necessary, through enforcement of safe work practices and safety and health rules?

2. Interviews

a. Are supervisors aware of their responsibilities?
b. Do employees say that supervisors are carrying out these duties?

3. Site conditions and root causes

a. Has supervisor lack of understanding of safety and health responsibilities played a part in creating hazardous activities or conditions?

C. Ensuring that managers understand their safety and health responsibilities

1. Documentation

a. Do training plans and training records indicate that training for safety and health responsibilities is part of the training that managers receive and that all line managers have received it?

2. Interviews

a. Do employees indicate that managers know and carry out their safety and health responsibilities?

3. Site conditions and root causes of hazards

   a. Has managerial lack of understanding of safety and health responsibilities played a part in creating hazardous activities or conditions?

SAMPLE SELF-EVALUATION CHECK LIST

Instructions: Fill out the last three columns after judging the *effectiveness* of each part of your safety and health program. If the part of your program indicated in the first column is *fully* effective, enter "yes" in the second column. If not, enter "no," and explain the deficiencies in the third column marked "Comments." If specific improvements are planned, enter the date they are *completed* in the last column.

INDICATORS	YES/NO	COMMENTS	DATE COMPLETED
Management Commitment and Employee Commitment			
Policies and objectives are established and communicated to all employees.			
Top management is visibly involved in safety and health.			
Employees are involved in problem finding and solving.			
Safety and health responsibility is clearly assigned.			

Adequate authority and resources are given to those with responsibility.		
Managers, Supervisors, and employees are held accountable.		
Program operations are reviewed at least annually to evaluate success in meeting goals and objectives and for preparing new objectives.		
Worksite Analysis		
A baseline comprehensive survey of hazards has been conducted and updated within the past three years.		
Change analysis is done for every change of facility, equipment, process, or materials.		
Job hazard analysis is conducted on an ongoing basis.		

Self-inspections are conducted regularly by adequately trained personnel.		
Routine inspections are conducted by adequately trained personnel.		
Employees know how and whom to notify about hazards without fear of reprisal, and receive timely and appropriate responses.		

Accident Investigation

Accidents and near-miss incidents are investigated to identify all root causes and contributing causes and to prevent future occurrences.		

Recordkeeping

The OSHA 200 Log and other records are properly maintained and up to date. Injury and illness experience are tracked and reviewed on a regular basis.		

Hazard Prevention and Control

All identified hazards are prevented or controlled in the best manner feasible.		
Safe work procedures, based on job hazard analysis, have been established and reviewed on a regular basis.		
Supervision reinforces safe work through positive feedback and training.		
Enforcement of safe work procedures and safety and health rules is accomplished fairly and efficiently through a written disciplinary system that all employees understand.		
New and repeat hazards are identified and corrected in a timely manner.		

The facility and equipment are maintained regularly to prevent hazardous breakdowns.		
Arrangements with Occupational Health Specialists provide medical services, including health problem identification.		
Employees trained in first aid and CPR are available on every shift.		
Emergencies of all types have been prepared for.		
Exits, evacuation routes, and emergency telephone numbers are prominently displayed.		
Safety and Health Training		
Managers can explain their safety and health responsibilities.		

Supervisors can explain safety and health rules and procedures for hazard control, how to teach them to employees, and how they are enforced.		
Employees can explain how and why they do their job safely and healthfully.		
Employees use all required PPE.		
Employees can explain why PPE is used, how it is used, how it is maintained, and the limits of its protection.		

Sample from an Evaluation Instruction

Answer the questions in each section in narrative form, with complete sentences. You may use one sentence to answer more than one question if appropriate. Your answers should reflect completed judgments. Avoid use of terms that reflect incomplete judgments, such as "appears," "apparently," or "seems."

Section 2. Routine Inspections

1. List any objectives set that involve routine inspections.
2. Are inspections following the procedures set for them? If not, describe how they are different and how frequently this difference occurs.
3. If procedures are not being followed, discuss the reasons and whether the alternative activity meets the objective set for inspections.
4. If any specific objectives were set for inspections, were they met effectively? Explain.
5. Has inspection activity contributed to identifying and controlling hazards and potential hazards? Explain.
6. If not otherwise covered above, please answer the following and provide examples:
   a. Are inspections revealing hazards that are slipping out of their controls, such as guards removed from machines, housekeeping problems, or work procedures not being followed? If so, be sure to deal with this topic in the evaluation of hazard prevention and control.

b. Are inspections revealing new hazards that were been identified before? If so, be sure to deal with this topic under the evaluation of comprehensive surveys, change analysis, or routine hazard analysis.

c. Are inspections revealing repeated instances of the same problem? If so, are they the same problem uncorrected or the same problem recurring? If the former, deal with this problem under the evaluation of tracking of hazard correction or accountability, or both. If the latter, be sure to deal with the problem under hazard prevention and control.

7. List any recommendations for changes in routine inspection activity needed for next year. Make sure that the recommendations logically reflect the conclusions made above.

Note: The questions asked in this sample instruction are similar to those that should be asked for every element in your safety and health program if you are to achieve a thorough and meaningful evaluation.

Appendix 4: Process Hazard Analysis Worksheet

1. Name the process.

 ____

2. Give the date operations began. ____________________

3. Write a short description of the process.____________

 ____

 ____

 ____

 ____

4. Conduct an overview, write a narrative, and attach it to this form.

 Date completed ____________________

5. Prepare process flow chart and attach.

 Date completed ____________________

6. For each step in the process as shown on the flow chart, identify and record on the reverse side of this form each:

 a. hazardous substance
 b. piece of equipment that presents a hazard
 c. employee operation that may be hazardous

Appendix 5: Suggested Inspection Inventory

1. Environmental factors (such as illumination, dusts, gases, sprays, vapors, fumes, noise, and heat and cold)
2. Hazardous supplies and materials (such as explosives, flammables, acids, caustics, and toxic materials or by-products)
3. Production and related equipment (such as mills, shapers, presses, borers, lathes, grinders, and saws)
4. Power source equipment (such as steam and gas engines and electrical motors)
5. Electrical equipment (such as switches, fuses, breakers, outlets, cables, extension and fixture cords, ground connectors, and connections)
6. Hand tools (such as wrenches, screwdrivers, hammers, chisels, and power tools)

7. Personal protective equipment (such as hard hats, safety glasses, safety shoes, and respirators)
8. Personal service and first aid facilities (such as drinking fountains, wash basins, soap dispensers, safety showers, eyewash fountains, first aid supplies, and stretchers)
9. Fire protection and extinguishing equipment (such as alarms, water tanks, sprinklers, standpipes, extinguishers, hydrants, and hoses)
10. Walkways and roadways (such as ramps, docks, sidewalks, aisles, and vehicle ways)
11. Elevators, electric stairways, and manlifts
12. Working surfaces (such as ladders, scaffolds, catwalks, platforms, sling chairs, and desk heights)
13. Material handling equipment (such as cranes, dollies, conveyors, hoists, forklifts, chains, ropes, and slings)
14. Transportation equipment (such as automobiles, railroad cars, trucks, front-end loaders, helicopters, and motorized carts and buggies)
15. Warning and signaling devices (such as sirens, crossing and blinker lights, klaxons, and warning signs)
16. Containers (such as scrap bins, disposal receptacles, carboys, barrels, drums, gas cylinders, and solvent cans)
17. Storage facilities and areas, both indoor and outdoor (such as bins, racks, lockers, cabinets, shelves, tanks, and closets)
18. Structural openings (such as windows, doors, stairways, sumps, shafts, pits, and floor openings)

19. Buildings and structures (such as floors, roofs, walls, and fencing)

20. Miscellaneous

Source: National Safety Council/OSHA Training on Safety and Health Management

Appendix 6: Employee Report of Hazard(s)

Note: This form is provided for the assistance of the employee making the report and does not constitute the exclusive means for notification of hazards.

I believe that a condition or practice at the following location is a job safety or health hazard.

Is there an immediate threat of death or serious physical harm?

Yes _____ No _____

Provide information that will help locate the hazard(s) such as building or area of building or the supervisor's name.____

____

Describe briefly the hazard you believe to exist and the approximate number of employees exposed to it.________

____

____

____

If this hazard has been called to anyone's attention, as far as you know, please provide the name of the person or committee notified and the approximate date.__________

____

____

____

Signature (optional)_________________________________

Typed or printed name (optional)______________________

Date __________________

Management evaluation of reported hazard(s) __________
____
____

Final action taken ______________________________
____
____
____

All actions completed by __________(date) Initials _____

Appendix 7: Sample Form for Hazard Correction Tracking

Instructions: Under the column headed "System," note how the hazard was found, using "Insp." for inspections, "ERH" for employee reports of hazards, or "Acc." for accident/incident investigations.

Under the column headed "Hazard Description," take as many lines as you need to describe the hazard. Under the third column, provide the name of the person who has been assigned correction responsibility. Under the fourth column, list any interim action to correct the hazard and the date. Under the last column, list the completed corrective action and the date that the final corrective action was taken.

HAZARD CORRECTION TRACKING WORKSHEET

System	Hazard Description	Assigned to	Interim Corrective Action/Date	Final Corrective Action/Date

Appendix 8: Preparing a Disciplinary System Worksheet

The type and severity of disciplinary action should be appropriate for the severity and frequency of the violation. Below is a series of questions designed to help you develop a disciplinary system that best meets the needs of your workplace. You may have already addressed the first two areas when you developed safe work procedures for different jobs at your workplace. If you have not yet developed these procedures, you will want to do so before developing a disciplinary system. Other workplace problems, such as attendance and attitude, are equally important but are not addressed here.

1. *Operations*: What key operation or operations occur at your workplace? What equipment is used, and by whom? What materials are used, and by whom? Are there any hazards associated with the use of the equipment or the materials?

2. *Procedures*: What are the key types of jobs at your workplace? What do most people do in the course of their work? What is the most efficient way for them to perform their jobs? What is the safest way for them to perform their jobs?

3. *Problems.* What would happen if a job or a procedure were not done safely? Exactly what would happen if an employee performed in an unsafe or unhealthful manner? What would happen if all employees did the same thing? How serious would the consequences be? Would the unsafe action or behavior affect just one employee, or all employees?

4. *Correction.* For each type of safety and health violation you have identified, what kind of corrective action seems appropriate? What would you do for a second offense, or for repeated violations of the same rule? Should warnings be oral or written? How long a suspension is warranted for what type of violation? Are there any actions that should automatically result in termination?

For this last stage of developing your disciplinary system, you may find it helpful to develop a grid to identify corrective actions for different types of violations and their frequencies. In the example following, a few types of safety problems are listed down the left side; across the top are indications of their frequency. You can fill in each box with the type of corrective action (i.e., oral warning, written warning, re-instruction, suspension, termination) that you consider appropriate.

	First Offense	Second Offense	Repeated Violations
Unsafe work habits	________ ________	________ ________	________ ________
Refusal to follow safety instructions	________ ________ ________	________ ________ ________	________ ________ ________
Unsafe actions that jeopardize self and & others	________ ________ ________ ________	________ ________ ________ ________	________ ________ ________ ________

Examples of Several-Step Disciplinary Systems

Most disciplinary systems use corrective procedures that involve three, four, or five steps. These are described briefly below.

Three-Step System

First violation:	Written warning; copies to employee and employee's personnel file
Second violation:	Written warning; suspension for half or full day without pay
Third violation:	Written report for file and immediate termination

Four-Step System

First violation:	Oral warning; notation for employee's personnel file
Second violation:	Written warning; copy for employee's personnel office file

Third violation:	Written warning; one-day suspension without pay
Fourth violation:	Written warning and one-week suspension, or termination if warranted

Five-Step System

First violation:	Instruction to discuss violation, proper procedures, and the hazards they control; notation for supervisor's file
Second violation:	Re-instruction, with notation in employee's personnel file
Third violation:	Written warning describing the violation and actions that will be taken if it reoccurs
Fourth violation:	Final warning; may include suspension
Fifth violation:	Termination

Note: The use of these corrective procedures obviously will vary with the nature of the problem and the frequency with which it occurs. Violations of rules are generally considered more important than other employee problems, but each requires correction. Keep in mind—and tell your employees—that your goal is to control unsafe *acts and conditions* in order to prevent *accidents.*

Index